Synchronous Precharge Logic

Synchronous Precharge Logic

Marek Smoszna

AMSTERDAM • BOSTON • HEIDELBERG • LONDON
NEW YORK • OXFORD • PARIS • SAN DIEGO
SAN FRANCISCO • SINGAPORE • SYDNEY • TOKYO
Morgan Kaufmann is an imprint of Elsevier

ELSEVIER

MORGAN KAUFMANN

Acquiring Edior: Todd Green
Editorial Project Manager: Robyn Day
Project Manager: Danielle S. Miller

Morgan Kaufmann is an imprint of Elsevier
32 Jamestown Road, London NW1 7BY
225 Wyman Street, Waltham, MA 02451, USA

Notices
Knowledge and best practice in this field are constantly changing. As new research and experience broaden our understanding, changes in research methods or professional practices, may become necessary. Practitioners and researchers must always rely on their own experience and knowledge in evaluating and using any information or methods described herein. In using such information or methods they should be mindful of their own safety and the safety of others, including parties for whom they have a professional responsibility.

To the fullest extent of the law, neither the Publisher nor the authors, contributors, or editors, assume any liability for any injury and/or damage to persons or property as a matter of products liability, negligence or otherwise, or from any use or operation of any methods, products, instructions, or ideas contained in the material herein.

Library of Congress Cataloging-in-Publication Data
Application submitted

British Library Cataloguing-in-Publication Data
A catalogue record for this book is available from the British Library

ISBN: 978-0-12-398527-9

For information on all Elsevier publications
visit our website at elsevierdirect.com

This book has been manufactured using Print On Demand technology. Each copy is produced to order and is limited to black ink. The online version of this book will show color figures where appropriate.

Working together to grow libraries in developing countries

www.elsevier.com | www.bookaid.org | www.sabre.org

ELSEVIER   BOOK AID International   Sabre Foundation

Transferred to Digital Printing in 2012

Dedicated to my parents Maria and Jerzy Smoszna, to my wife Justyna, and to my wonderful children, Zuzanna and Michał, whose exuberance and joy always help me remember what is really important.

Contents

List of figures

List of tables

About the author

Marek Smoszna is a memory design engineer at Nvidia Corporation. He was born in Poland in 1972. He received his BS and MS in electrical engineering from Rensselaer Polytechnic Institute in 1995 and 1996, respectively. He subsequently took several analog and digital integrated circuit design courses at Stanford University. Smoszna holds two patents in the area of memory design, with several other patent applications filed. His interests include high-speed circuit and memory design. When not doing circuit design, he can be found with his children at the beach.

1 Precharge Logic Basics

1.1 Introduction

The purpose of this book is to describe the issues involved in precharge circuit design and to establish design guidelines that will minimize the design risk. Anyone designing dynamic circuits or full swing memories should read this book.

There are many logic families utilizing the metal oxide semiconductor (MOS) transistor. Static complementary metal oxide semiconductor (CMOS) is the most widely used, partly because it is safe. "Safe" means that it will almost always work without any special considerations. However, sometimes static logic just is not good enough. In advanced very large scale integration (VLSI) systems such as microprocessors, there arise "critical" logic paths where static logic is just too slow to meet the timing constraints. There are faster logic families, and precharge (dynamic) logic [1] is one of them. Precharge logic also has other advantages over static logic [37]. Of course, there are also disadvantages, and design with precharge logic is more challenging to ensure proper circuit operation.

This book covers synchronous (clocked) precharge logic. "Clocked" simply means that the clock is used to (indirectly) precharge the outputs of the logic gates. There exists also asynchronous precharge logic, but the clocked version is simpler and more popular. This type of logic is widely called "dynamic" logic in the industry, where "dynamic" simply means relying on charge storage, and "static" means "always driven." But because "keepers" can be added to the dynamic nodes, this type of logic can be dynamic or static. Thus, "precharge logic" appears to be a more appropriate name. Keepers will be explained in detail in Section 1.7 and in Chapter 3.

1.2 What Is Precharge Logic?

A basic precharged NAND gate is shown in Figure 1.1. The logic gate consists of an N-channel metal oxide semiconductor (NMOS) tree of transistors known as the "pulldown stack." The output is precharged HIGH by the P-channel metal oxide semiconductor (PMOS) device when the clock is LOW. The output is conditionally discharged when the clock turns on the NMOS device connected to ground. This is known as the evaluation device or "footer." Discharge of the output happens when the clock is HIGH and the inputs create a path in the NMOS tree from the output node to the bottom NMOS device. The clock can be replaced with a signal named "/precharge," as in the general case the precharge/evaluate functions can be

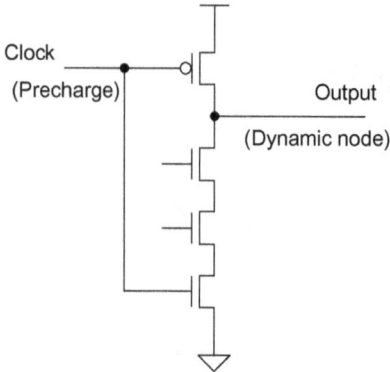

Figure 1.1 Basic precharged gate (NAND) with dynamic output.

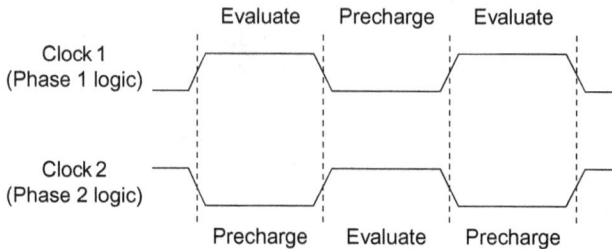

Figure 1.2 Clock phase definition.

controlled with any signal. An example would be an asynchronous system where there are no clocks. In any case, please note that we precharge HIGH and evaluate LOW because of the good efficiency of the NMOS devices. Predischarging LOW and evaluating HIGH would not be as efficient.

Some common terms used are as follows (Figure 1.2):

the *precharge phase* is when the clock input to a circuit is low,
the *evaluation phase* is when the clock input to a circuit is high.

1.3   Why Is it Faster than Static Logic?

Precharge logic gets its speed from the fact that the pullup tree (single P-channel field effect transistor (PFET)) and the pulldown tree are never on at the same time. When a static gate switches, there is an overlap time when the two trees are "fighting." Precharge logic separates the pullup and pulldown times, as can be seen in Figure 1.3. If we want to discharge the output (pull it down), then the only current flowing in the NMOS tree is from the charge stored on the output node. In static logic, there is also a short circuit (crowbar) current that flows from V_{dd} to V_{ss}, and so the peak transition current in static logic is higher than in precharge logic. Because the crowbar current exists directly between V_{dd} and V_{ss}, it can be seen that this current does no logical work and is therefore wasted power. More importantly,

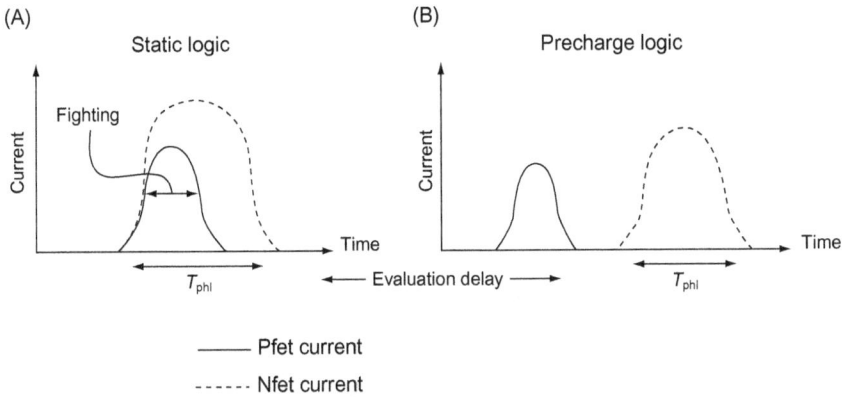

Figure 1.3 (A) Static transition HIGH to LOW; (B) precharge logic precharge and discharge.

as the current in the metal oxide semiconductor field effect transistor (MOSFET) is not solely from the charge stored in the load capacitance, it takes longer to discharge the load in static logic.

Another factor in the precharge logic speedup is the fact that the dynamic node starts to pull down when the input reaches the transistor threshold voltage V_t. This is sooner than in the static gate, which begins switching when the input reaches roughly $V_{dd}/2$.

So the pulldown speed is improved considerably, even though there is typically an extra NMOS device in series with the NMOS logic tree. This is of course the clock-controlled footer, connected to V_{ss}. The pullup time experiences the biggest improvement in speed because there is only one pullup device (as opposed to a series P stack). But because this is the precharge time, it is typically not critical and so timing can be adjusted depending on how one sizes the device.

One may ask how this logic can be faster if a whole half of a clock cycle is dedicated to precharging the output. The answer is that the whole picture must be considered, meaning the logic and latches/registers separating the logic stages. In a typical two-phase clocking scheme, each logic stage is only given one half of a clock cycle to evaluate anyway. So while one block is evaluating, another is precharging and vice versa, as can be seen in Figure 1.4 [7].

There are other ways of using precharge logic, as described in Ref. [3]. In this case, static and precharge logics are interleaved and static logic propagates signals while precharge logic is being precharged. In this way, there is still signal propagation on every phase. This approach is a compromise between using static and precharge logics.

There are times when precharge logic is not faster than static. With latch-based clocking schemes, static logic can take advantage of "time borrowing" [7], which may result in better timing than precharge logic. Time borrowing simply means that signals can flow to the next block of logic without having to wait for a clock edge. Thus, the next block starts evaluating early, in a way "borrowing" time from the current block. But precharge gates can sometimes take advantage of time borrowing.

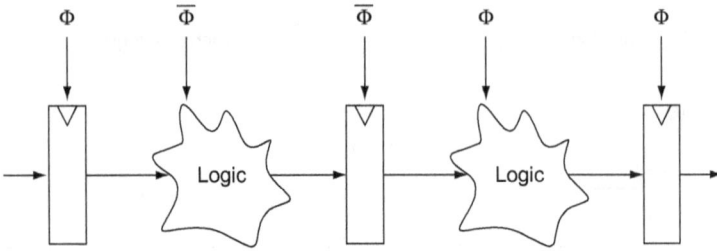

Figure 1.4 Precharge logic with latches.

If the logic stages on both sides of a latch consist of precharge logic, then time borrowing is possible [7]. This is done by overlapping the evaluation phases of both stages. We will discuss this in detail in Chapter 2. Furthermore, the use of a dual monotonic (dual rail, precharged) latch also removes the synchronization point that is created between latches and precharge logic blocks [4]. It is therefore clear that there are numerous ways to overcome synchronization point problems in precharge logic. As with other logic families, a logic chain is limited by the duration of the evaluation phase. All outputs must be stored at the point that the latch on the output closes.

Let us now summarize the factors that impact the speed of precharge logic:

- no fighting between PMOS and NMOS trees,
- evaluation starts as soon as we reach V_{tn} of the NMOS device,
- lower input capacitance for the same output current,
- inverting static gate can be skewed in favor of the critical edge.

1.4   Advantages of Precharge Logic

There are a number of advantages to using precharge logic. As mentioned above, precharge logic is faster than static logic due to several factors. In addition, the precharge logic gates are smaller in physical size because there is only a single PMOS pullup device. Of course, this only holds true when implementing Boolean functions that are more complicated than a simple inverter. Not having to implement a full PMOS tree can thus result in large savings in areas in logic gates with a large fan. The reason for this is that the carrier mobility in PMOS devices is lower than that of NMOS devices. This condition requires the use of significantly larger PMOS devices to achieve the same resulting conductivity as that of the smaller NMOS devices. In addition to the area penalty for the larger PMOS devices, a performance penalty will be paid for their use due to self-loading and input loading. Self-loading comes from the fact that any increase in the node capacitance on the output of a logic gate will reduce its performance. So as we increase the width of a transistor, its associated parasitic capacitance increases as well. The reduction to self-loading (as compared to static circuits) allows the efficient construction of gates such as an 8 input multiplexer or NOR gate.

Because the precharge logic gates are smaller, the gates that drive them can also be smaller. Naturally, this is because the precharged gates present a smaller load capacitance to their drivers. The other way to look at this is that the driving gates can be faster, if not reduced in size. Not reducing the driver size may not be the best choice; however, a fanout of 3 to 4 is known to be optimum for most applications. Furthermore, larger gates may still not be able to compensate for wire RC, as shown in Figure 1.5. If there are two resistors in series, the total resistance can never be smaller than any one of the resistors.

So far it was established that precharged gates are faster because there is no fighting between the pullup and pulldown networks. It also was pointed out that the pulldown time is not improved as much as the pullup time because there is still a full NMOS tree of transistors and there is an additional NMOS device placed in series with the tree. But as we will learn in Section 1.6, just about all precharged gates are followed by a static inverter, as shown in Figure 1.6, which happens to be a NOR gate. While this will be discussed in detail in Section 1.6, it is worth pointing out that with the static inverter, the whole logic gate can evaluate faster, even with the additional propagation delay from the inverter. This is due to the fact that now there are not any series transistor stacks driving the next gate, there is only an inverter with a single PMOS and NMOS device, driving the load with an inverter is potentially the fastest one can get in a given technology. In order for the inverter to provide such improvement in speed, its PMOS device should be sized much larger than its NMOS device. However, this severely reduces the noise margin. In

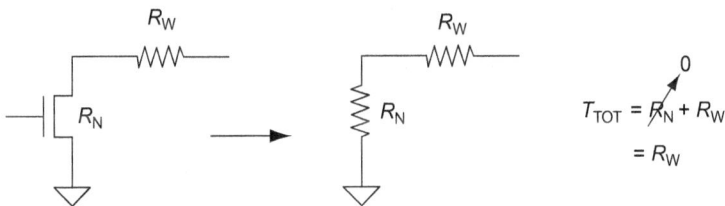

Figure 1.5 Transistor driving a wire load.

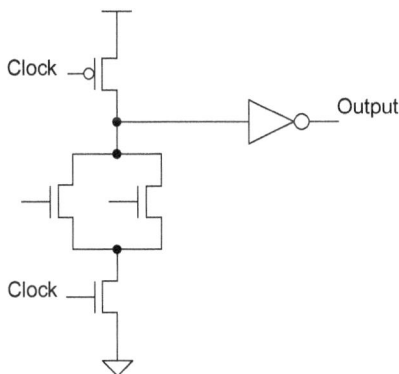

Figure 1.6 Precharged NOR gate with a static inverter becomes an OR gate.

fact, some people increase the NMOS device to improve the noise margin at the expense of speed. If a circuit is designed to be fast but it does not work on actual silicon, then it is useless. Improvements to the noise margin will be discussed in Chapter 4.

Precharge logic gates are inherently less noisy because there is no short circuit (crowbar) current flowing from power to ground during each transition. The crowbar current is such that it has a large initial spike and a gradual rolloff, rather than a more moderate and constant level. This spike of crowbar current has the tendency to cause local power supply deviations. This is normally the source of ΔI noise. As less current flows, there is also less power dissipation in the logic gate. There is also less power dissipation at the input to the logic gate because the input capacitance is smaller. However, because the clock must be routed to many more places on the chip, there is increased power dissipation in the clock wires. Furthermore, if an input to a precharged gate is not switching, the output might switch with the clock, which increases overall power dissipation. Therefore, it is not clear what the impact on power dissipation is, and this has to be addressed on a case-by-case basis.

Another advantage of precharge logic design is that these circuits allow the designer to optimize the transistors for one edge of interest. This is in direct contrast to static circuits that will need to make their rising and falling output edge rates nearly equal.

Yet another advantage to precharge logic design is that these circuits can be expanded to include a latch mechanism in the logic circuitry without significantly slowing down the circuit. This is discussed in detail in Chapter 2 and also in Chapter 7.

A final advantage to precharge (domino) gates is that any resulting circuit will be glitch-free by construction. This is due to its single transition nature. As we will learn in Section 1.6, the only transition that a domino circuit can make during evaluation is that of a zero to one transition, on the output of the circuit.

It is now worthwhile to summarize the advantages and also mention a few disadvantages of precharge logic.

Advantages:

- Faster switching
- Less noise produced
- Less power dissipation (potentially)
- Smaller gate input capacitance
- Smaller layout area
- Optimized for one edge of interest
- Integrated latch mechanism possible
- Glitch-free by construction (domino)
- Great for high fan-in gates.

Disadvantages:

- Lower noise margin
- Difficulty with "time borrowing"

- Lack of inversions (domino)
- The need to route the clock to all gates
- Tricky design (charge sharing and leakage on dynamic nodes)
- More power dissipation (potentially)
- Difficult to interface with, require monotonic signals (domino)
- Require a precharge phase to prepare them for the next logic evaluation
- Minimum frequency of operation—cannot hold state in static mode.

In addition, it should be noted that precharge logic is a ratioless logic family, meaning that the slope of the transfer characteristic is not dependent on the ratio of the sizes of the PMOS and NMOS transistors.

1.5 What About Using Other Transistors?

Precharge logic can only be realized with MOS transistors. In order for a node to retain its charge for a reasonable amount of time, all leakage currents must be very small. When using MOS transistors, the leakage currents can be indeed small (though in modern processes leakage has increased considerably). In any case, compared to other transistor types, the enhancement mode MOS transistor can be fully turned off, thus lowering leakage. Other technologies based on devices such as bipolar junction transistors (BJTs), heterojunction bipolar transistors (HBTs), metal semiconductor field effect transistors (MESFETs), high electron mobility transistors (HEMTs), depletion mode MOSFETs, etc., in general, all have leakage currents too high to retain charge on the dynamic node. Furthermore, the MOS transistors have a good insulator between the gate and the channel, i.e., the gate oxide. This also helps to retain the charge, because the charge does not really get stored on the output of a logic gate, but on the next gate's input. Dynamic storage is a major advantage of MOS devices, though in modern technologies the oxide is very thin and there is more leakage through the oxide. To keep the drain current reasonable and also to prevent the oxide from getting too thin, a high "k" dielectric is used in modern processes [26].

1.6 Domino Logic

Domino logic is a special version of precharge logic, which has a certain restriction on the signals flowing from logic gate to logic gate. The restriction is that all signals on the inputs of a precharged logic gate, in the evaluation phase, can only transition from LOW to HIGH, or remain LOW. These types of signals are referred to as *monotonically rising*, because they can only make one transition, from LOW to HIGH (in the evaluation phase). The word "monotonic" means "changing in one direction only" [4].

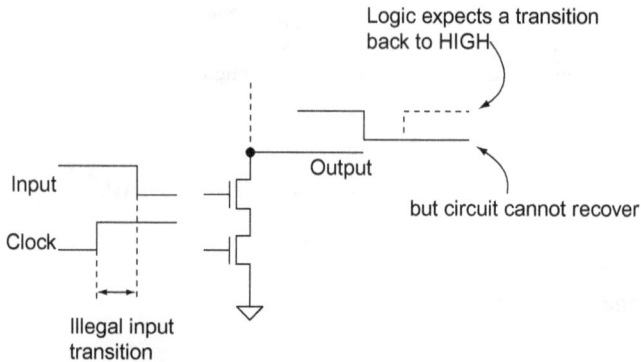

Figure 1.7 Discharge of an output node due to an illegal input transition in an evaluation phase.

1.6.1 Need for Monotonic Signals

Precharged gates such as the one shown in Figure 1.1 cannot be cascaded in a simple manner. These gates require that all inputs be stable before the start of an evaluation phase of a gate, or be monotonically rising after the evaluation phase has started. So if we had a block of logic, each logic gate would have to have a separate clock to ensure that each gate is precharged when its inputs are unstable. This is ridiculous both because of the enormously complicated clocking scheme it would require and because propagation delays vary from logic block to logic block. But if this condition is not met, simple precharged logic gates cannot be cascaded [6].

If precharged logic gates are to be cascaded, limitations must be put on the signals that flow from gate to gate. If a gate started evaluating a signal that bounced around between power and ground, the precharged gate would be discharged without possible recovery. Or, if a signal started out HIGH and settled down to a LOW, then the output we want is a HIGH. But the output was discharged by the input starting HIGH (assuming evaluation phase, of course), creating a path from the output to V_{ss}, as can be seen in Figure 1.7. Once the input settles to a LOW, there is nothing to restore the output and it will remain LOW. So this is why all signals should start LOW, and at most make one transition from LOW to HIGH [6].

1.6.2 Domino Logic Gates

The modification to existing precharged logic gates is simple. All that is needed is the addition of a static inverting gate on the output of the precharged logic gate [12]. This is shown in Figure 1.8. A domino gate actually refers to two stages, rather than a single gate. In any case, in this way a block of logic gates can be precharged with the same clock. So, when a logic gate is precharged, the dynamic node is precharged HIGH, but the output of the static gate is LOW. Thus, all inputs to precharged gates start out LOW in an evaluation phase, as desired. Another way to state this is that

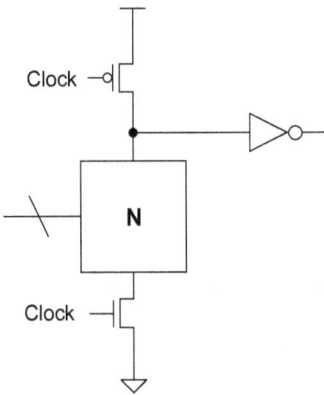

Figure 1.8 Standard domino gate.

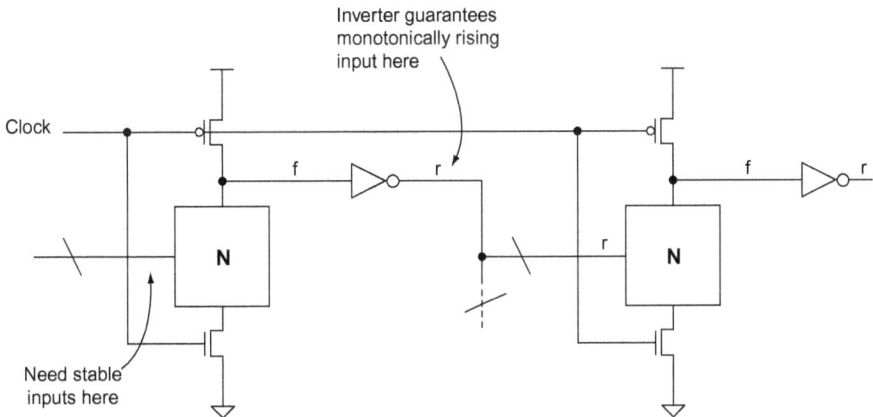

Figure 1.9 Chain of two domino logic gates.

the dynamic node is monotonically falling, and the output of the static inverting gate is monotonically rising. So by combining each precharged gate with a static inverting gate, the output is a valid signal into another precharged logic gate [4].

A queue of domino blocks can fall, but cannot get up by itself. Similarly, a precharged logic gate's output can only fall (NMOS) when it evaluates [4]. So a chain of gates looks like a queue of domino blocks. In Figure 1.9, "r" means signals are monotonically rising and "f" means signals are monotonically falling. The output of a domino gate starts LOW. It either stays LOW, or it raises HIGH. If it does not change, then the next domino gate is unaffected. But if it rises, then the next domino gate might transition its output from LOW to HIGH, depending on other inputs to that domino gate. As already discussed above, the static inverting gate (inverter in most cases) improves the noise margin and improves speed, because one device (as opposed to a series stack of transistors) drives the next gate.

The problem with domino logic is that it is noninverting. Thus, not all functions can be implemented with domino logic.

1.7   Keepers: Improving the Charge Storage

Precharge logic is in many cases dynamic logic, in the sense that its outputs are dynamic nodes. Sometimes there may be problems with charge storage due to excessive leakage in the circuit. "Keepers" may be used to replenish the lost charge. A keeper is a small (weak) PMOS device connected to the dynamic node and driven by a static inverter, as shown in Figure 1.10. This circuit restores the charge lost on the dynamic node.

Another example of using a keeper is shown in Figure 1.11. Here, a keeper is used on a Svensson style latch [5]. The dynamic node is the node in the middle of the circuit, marked "x." The reason for the node being dynamic is due to the fact that when the latch is closed (enable is LOW) and the input is HIGH, node "x" is floating. We only care about the case when this node is HIGH, because if it is LOW and drifts up in potential, there will not be any unwanted transitions.

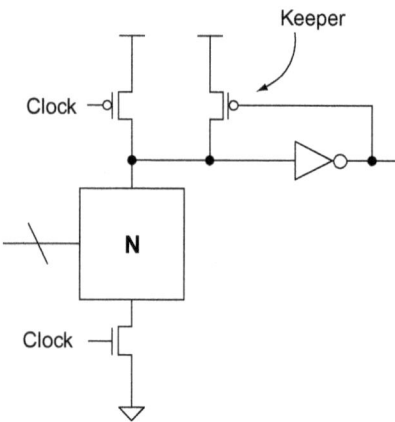

Figure 1.10 Use of a "Keeper" to maintain charge on the dynamic node.

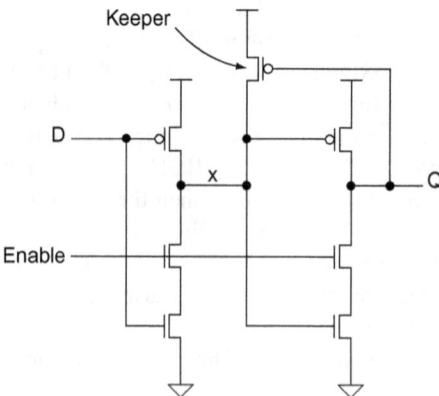

Figure 1.11 Svensson style latch with a "Keeper."

Because the second stage of the latch is inverting, no static inverter is necessary and the keeper can be included, as shown in Figure 1.11.

A keeper is always weak so that it does not slow down the conditional discharge of the dynamic node. Because it is weak, it cannot prevent instantaneous events from switching the output. An instantaneous event could be a large spike due to capacitive coupling or charge sharing (to be discussed in Chapter 4). In these situations, the keeper is too slow to start recovery of the dynamic node quickly enough for the logic gate not to make an unwanted transition.

Keepers transform dynamic nodes into static or weakly dynamic ones. Thus, the disadvantage of using them is that they fight the NMOS tree when it attempts to discharge the dynamic node. One must keep in mind that keepers must be sized correctly, and this will be discussed in Chapter 3. So the keepers do help with leakage problems and prevent "small" spikes from discharging the dynamic node. This is important because subsequent "small" spikes could accumulate degradation in the stored charge. Replacing the charge lost due to leakage is important if the circuit is to work at low frequencies (e.g., during chip testing). In this case, the dynamic node has to store charge for relatively long periods of time. So in general, most precharge circuits do use keepers [4]. Keepers do little to aid high-frequency operation, because the device does not have enough time to operate [6]. To summarize, keepers should be used if there is significant leakage (such as in most modern processes), coupling, and/or low-frequency operation.

As a last comment, it should be noted that PMOS keepers are most popular. But this only helps to maintain a HIGH value on the dynamic node. Sometimes there may be problems maintaining a LOW value after the gate has evaluated. If for some reason there is too high a voltage drop across a series transistor stack, the resulting V_{ol} on the dynamic node may be too high. Also, NMOS keepers may be necessary under special conditions when an evaluation phase must be maintained for a long time, and the inputs have gone LOW. This may happen during a scan operation while testing the chip. In such cases, an NMOS keeper is added (sometimes with its own evaluation device), as shown in Figure 1.12. In this case, the NMOS keeper acts to prevent excessive PMOS leakage from V_{dd}.

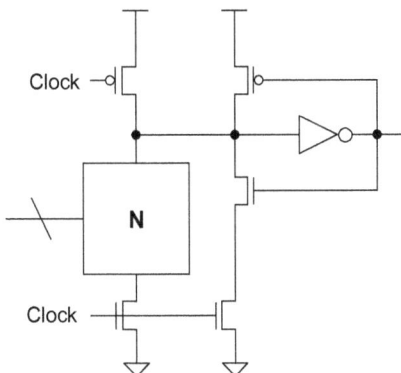

Figure 1.12 Use of both PMOS and NMOS Keepers.

1.8 Final Comments

Precharge logic designers face a number of questions when selecting their circuit topology. How many stages are best? Should the static gates be inverters or should they perform logic? How should they size the precharge devices and keepers? What is the benefit of removing the clocked evaluation device? All these questions will be answered in later chapters.

2 Timing

Clocking of precharge circuits requires the use of multiple clock phases. In the simplest case, two-clock phases could be used and this is often referred to as "traditional two-phase" design. In a more general case, one could use more than one-clock phase, such as four-clock phases [6], but this is beyond the scope of this book.

In two-phase clocking, we need to have two clocks. The chip can generate two clocks at the phase locked loop (PLL) and distribute both clock networks to all chip components. This requires quite a lot of overhead, so in many cases, designers choose to route just one clock and generate the complement at a more local level.

Let us now establish a naming convention for our clocks. The first phase will be known as "phase 1" and the second phase will be known as "phase 2" (Figure 2.1).

We need to partition the logic within a cycle into two groups, one series of gates for each of the two-clock phases. To maximize the amount of logic that can be done within each clock phase, the result from the last logic gate in each phase must arrive very close to the end of that phase. This is to minimize the time wasted waiting for the next clock signal. Time borrowing cannot be accomplished in a two-phase domino.

2.1 Clock Skew Penalty

A clock skew penalty between the two logic blocks exists because the rising edge of clock2 may be generated from a different clock header than the falling edge of clock1 (Figure 2.2). Because the two clock headers will have different delays in their clock distribution networks, we have clock *skew*. Furthermore, we will also see variations in the duty cycle, which are due to variations in the PLL and the clock tree that generates our clocks. This is referred to as clock *jitter*. Together, skew and jitter are referred to as the *clock uncertainty*:

$$\text{clock uncertainty} = \text{clock skew} + \text{clock jitter} \qquad (2.1)$$

Given the clock uncertainty, it is possible that the falling edge of clock1 will be early and the rising edge of clock2 will be late. In this situation, the circuit cannot do any work during the time between the two phases. The signal is waiting for the clock2 rising edge to arrive. Therefore, we must add this clock blocking time to our delay, and we refer to it as the "clock skew penalty."

Synchronous Precharge Logic.

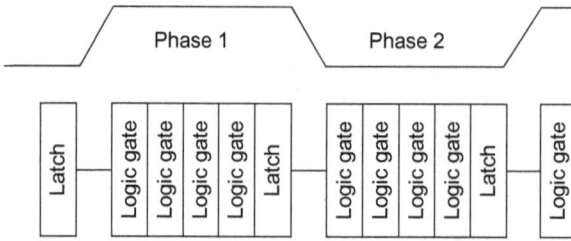

Figure 2.1 Diagram of clock phases.

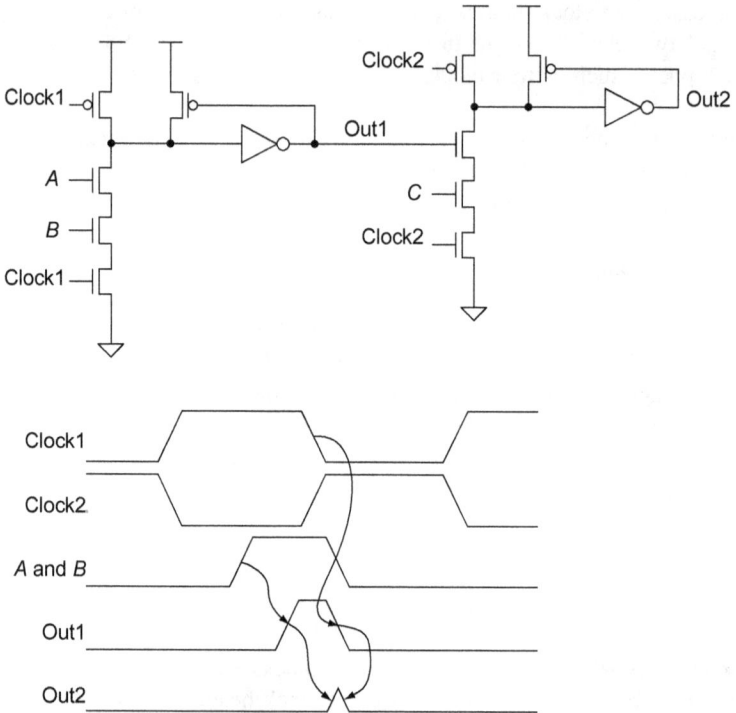

Figure 2.2 Hold-time failure at phase boundary.

As we will see in a later section, the clock skew penalty can be avoided if we provide an overlap of clock1 and clock2. For us to avoid clock blocking entirely, the overlap between the two clock signals must be greater than the total clock uncertainty between the two clocks.

2.2   Hold-Time Problem

When nonoverlapping clocks are used in dynamic design, a potential hold-time problem exists at the phase boundaries, as seen below:

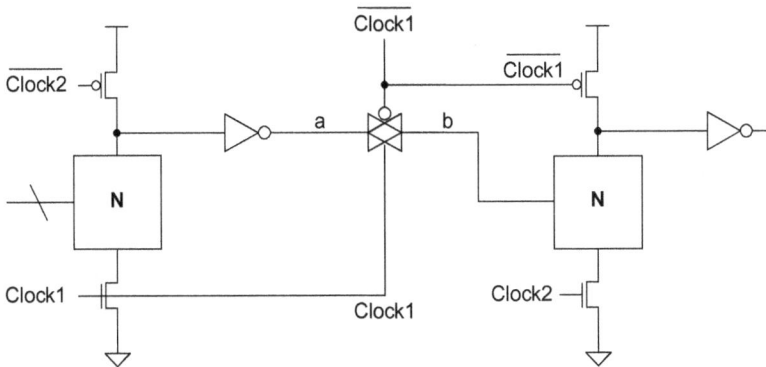

Figure 2.3 Nonoverlapping clock circuit diagram.

In this example, out1 evaluates to one at the end of phase 1 and precharges to zero at the beginning of phase 2. If out1 had held its result longer, out2 could have properly evaluated. So the problem is that precharge comes too early for the amount of logic in phase 1. Even if we had less logic, and out1 had held its result, it would be necessary to prove that the hold time was sufficient to guarantee proper evaluation of out2 before out1 precharged.

This hold-time problem is easily solved by the use of a latch on the output of a precharge gate, as can be seen in Figure 2.3.

2.3 Nonoverlapping Clocks

Two-phase nonoverlapping clocks are extra conservative. The clocks would be connected to the logic and latches as shown in Figure 2.3. The first stage is precharged during clock2 and evaluated during clock1. Similarly, the second stage is precharged during clock1 and evaluated during clock2. The point is that the output of each stage settles some time after the evaluation clock goes HIGH and does not change again until the precharge is asserted, when the other clock goes HIGH.

The timing for this scenario is shown in Figure 2.4. The nonoverlap in the clocks gets rid of any potential race between precharging and the latch, but we lose

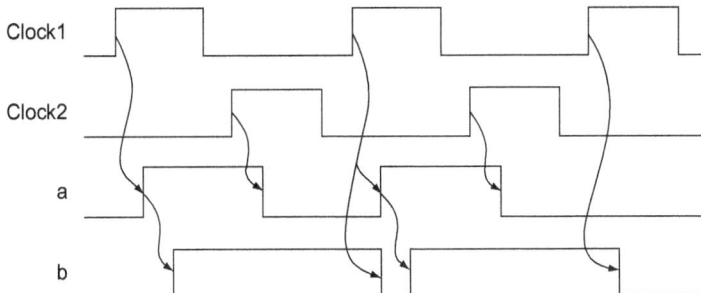

Figure 2.4 Nonoverlapping clock timing.

performance because not only do we still have no time borrowing but we also must pay the penalty for the nonoverlap time. Clock uncertainty adds to the timing penalty.

2.4   A Better Latch

Thus, nonoverlapping clocks can be used to solve the race problem between precharge and the latch. A better way to solve this problem is to use a clocked set−reset (CSR) latch in place of the output inverter in the last precharge gate. This approach is shown in Figure 2.5. Now the node QB is set LOW by the dynamic net falling. So the latch is already "holding" the state when precharge comes (clock goes LOW, dynamic net goes HIGH).

As we will see in Chapter 7, CSR latches have several interesting features. One of these features is the fact that there is no setup requirement. If we pull down the internal dynamic node, we will hold the data on the output properly, regardless of what the clock is doing. A second advantage is that the output is staticized, meaning it is not affected by precharge. The staticized result will be held through the entire next phase of the clock, thereby meeting all hold-time requirements for the first gate in the next phase.

The use of this structure brings an extra complication to the nature of the output signal. From Figure 2.6, it can be seen that the output of the CSR latch will appear

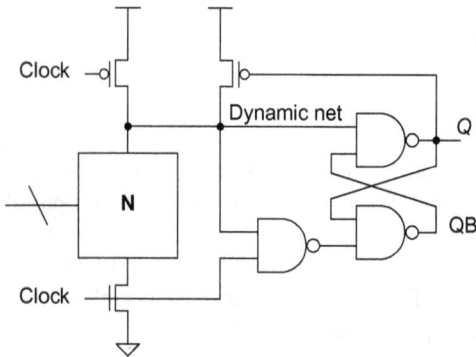

Figure 2.5 CSR latch placement at the end of precharge logic.

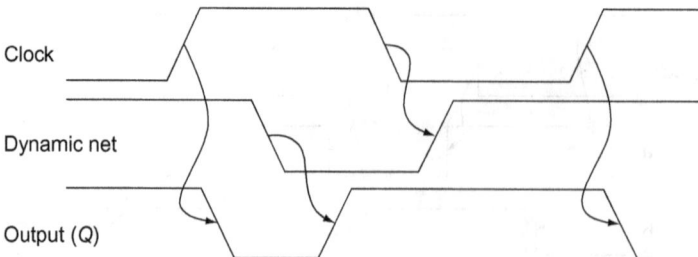

Figure 2.6 CSR latch signal timing.

to violate our previous rule against transitioning from one to zero, on the input of a precharge gate. So this CSR output is not monotonic.

In this instance, the output of the CSR latch has one characteristic that allows this falling transition to be safe. Namely, the gate after the CSR latch uses a different clock than that of the CSR latch. So, when the output of the CSR latch falls after the beginning of the CSR latch's evaluation cycle, the next precharge gate is in precharge. So the falling input does not affect it.

2.5   Input Setup Criteria

All falling inputs to the first precharge gate must reach 10% of V_{ss}, before the evaluation clock reaches 10%. If the input is rising, we can relax these criteria to 50% of V_{ss} for both the signal and clock. Figure 2.7 shows these concepts.

It is worth noting that the 10% signal level can be viewed as a source of noise. This kind of noise (incomplete precharge/discharge) typically is not part of any noise budget. However, there is a little bit of good news. This kind of incomplete discharge will add to other noise sources only if all signals (including the clock) experience the maximum amount of noise from other sources. Also, the noise limit would have to be the lowest possible, which would mean the smallest possible load on the dynamic net and widest possible noise coupling pulsewidth. The probability of all these things happening exactly at the same time is likely to be very low. This will be discussed in detail in Chapter 4.

2.6   Input Hold Criteria

All inputs to a gate should be present through the end of that gate's evaluation phase. In other words, all inputs to precharge circuits should maintain their logic one level through the end of the evaluation phase. This is to avoid an undriven low-dynamic node during the evaluate phase. Should this scenario occur, a full keeper must be used in the precharge gate. There is more information on keepers in Section 1.7 and in Chapter 5.

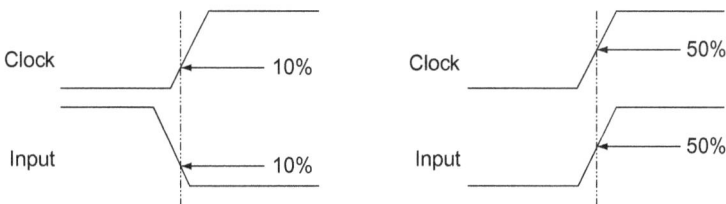

Figure 2.7 Input signal setup requirements.

2.7   Precharge Timing

Outputs of precharge circuits must fully reset (typically defined by the 10% point of the falling output) before the end of the precharge phase. This reset must happen properly at the far end of any transmitting wire. The reset also must take into account any clock uncertainty due to the distance to the far end of any output wire. Precharge also must account for the worst-case combination of inputs because the precharge operation will be slower if it must charge all the internal nodes in the evaluation stack (Figure 2.8).

2.8   Skew Tolerant Design

When precharge logic is pipelined in the same way as two-phase static logic is pipelined, they incur a significant amount of sequencing overhead from latch delay, clock uncertainty, and imbalanced logic (leading to dead time). By using overlapping clocks and eliminating the latches, we can hide all this sequencing overhead to achieve dramatic speedups. This is known as "skew tolerant domino" [30].

So in the case of two-domino logic chains, clocked by opposite-phase clocks, the need for the intermediate CSR latch can be eliminated altogether [30]. If there is a small amount of overlap between the clock phases, the output of a domino chain clocked by clock $\Phi1$ can directly drive the inputs of a domino chain clocked by clock $\Phi2$, without an intervening latch. See Figure 2.9, where the dark bold lines represent nominal clocks and the light lines represent the clock uncertainty.

So we need to overlap the clocks such that the next phase begins evaluation before the previous phase precharges. Once the first gate in the next phase has

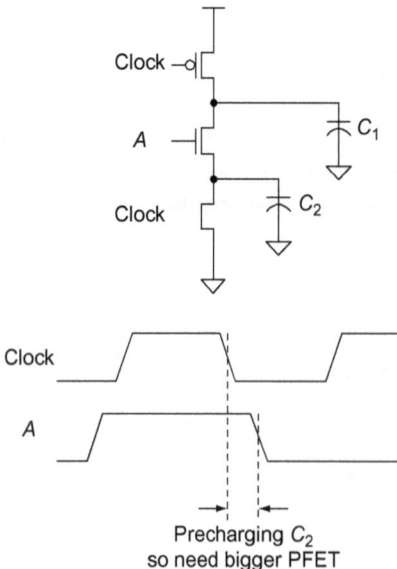

Figure 2.8 Precharging internal capacitances.

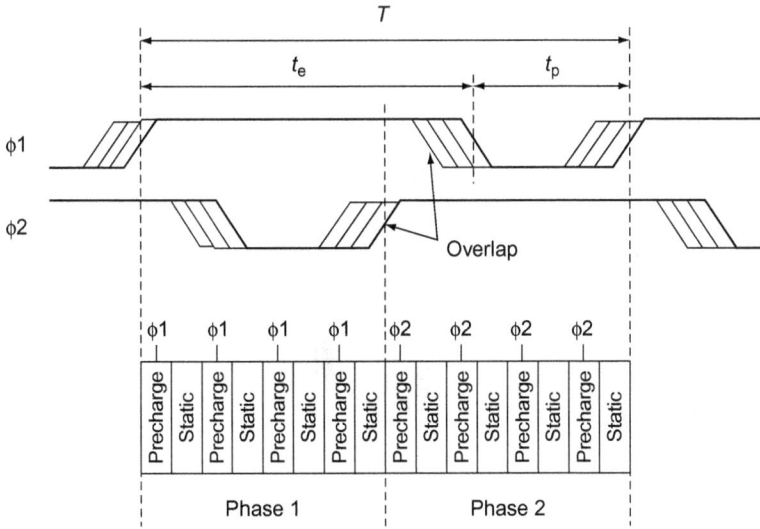

Figure 2.9 Two-phase overlapping clocks.

evaluated, the input is no longer needed and may fall low (predischarge) without impacting the next phase. Thus, latches are not needed as long as the clocks overlap sufficiently, as shown in Figure 2.9.

Clock uncertainty never impacts the critical path because the precharge gates are guaranteed to be in evaluation by the time critical data arrives. Furthermore, the gates do not precharge until the next gate consumes the result. Each phase is HIGH for an evaluation period t_e and LOW for a precharge period t_p, as shown in Figure 2.9.

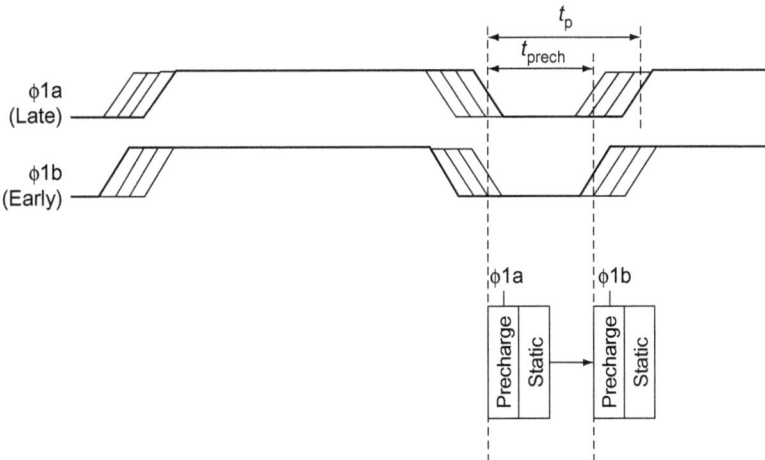

Figure 2.10 Precharge time constraint.

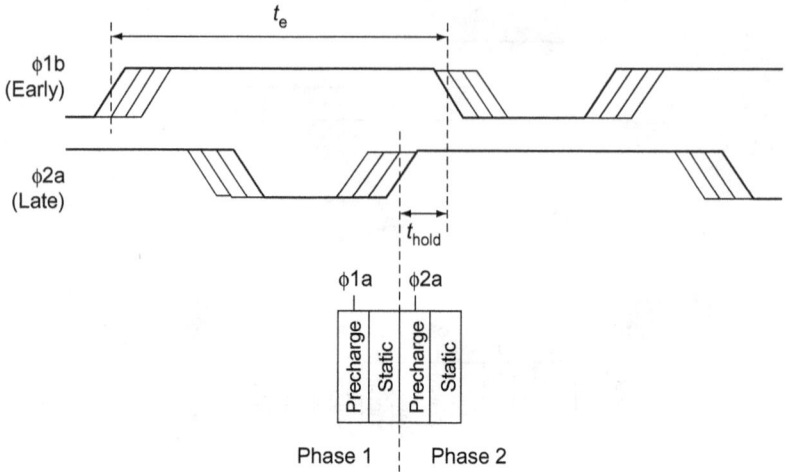

Figure 2.11 Evaluation time constraint.

Figure 2.10 shows the precharge time constraint. The worse case occurs when $\Phi 1a$ is skewed late and $\Phi 1b$ is skewed early, reducing the precharge window. From the diagram, we can write Eq. (2.2).

$$t_p \geq t_{prech} + t_{uncertainty} \tag{2.2}$$

Similarly, Figure 2.11 shows the constraint on evaluation time. It is set by the requirement that the logic remains in evaluation until the next phase logic consumes the result. The necessary overlap is called t_{hold} and the minimum nominal overlap in the phases is then $t_{hold} + t_{uncertainty}$. The evaluation time t_e is then given by Eq. (2.3).

$$t_e \geq T/2 + t_{hold} + t_{uncertainty} \tag{2.3}$$

3 Transistor Sizing

Proper precharge circuit sizing means adjusting the evaluation stack and output inverter sizes so that the gate is near minimum delay, while still allowing for proper precharge timing. Sizing precharge gates to achieve minimum delay will result in a circuit that is too large in area and power. Instead, the gate should be sized for near minimum delay.

As a rule of thumb, follow this procedure:

1. Estimate the lumped load.
2. Use fanout (FO) = 4 to do the sizing of the static gate.
3. Use FO = 2−2.75 for the precharge pulldown stack.
4. Run a simulation to check if edge rates are within range.
5. Optimize sizing for minimum delay while keeping edge rates in range.

3.1 Sizing the Pulldown Stack

In order to achieve reasonable sized gates and avoid electromigration (EM) problems, make sure that the pulldown edge rate is not too small. This also minimizes Miller backcoupling (source/drain to gate), which can distort the input edge. The worst case for this situation is when all stacks turn on (unless they are logically exclusive). It is best to have a fanout of 2.75 for footed precharge stacks and a fanout of 2 for footless precharge stacks [27].

Given the phase-based dependence of these circuits, the maximum edge rate should be about half that used for static gates. The worst case for this simulation is when just one stack turns on. As a rule of thumb, the maximum dynamic node fanout Nload:Nstack should be less than 12 in modern processes.

3.2 Sizing of the Output Inverter

Size the output inverter for reasonable fanout and keep the edge rates in range. As a rule of thumb, the output inverter fanout should be less than 12. The *beta ratio* is defined as the ratio of the P-channel field effect transistor (PFET) width to the N-channel field effect transistor (NFET) width, assuming both devices have the same channel length. Note that *mobility* is not taken into account. The output inverter beta ratio should be within the range of 1−4.

Values greater than 4 will create precharge problems and will increase the sensitivity of the gate. Using a small beta ratio will hurt the evaluation edge. In the event of considerable timing margin, it is better to keep the inverter beta above 3 and downsize the NMOS pulldown stack. If the fanout is greater than 5, the output driver and feedback inverter should be separate. The feedback inverter should have a beta = 1.5, if separate from the output driver.

3.3   Logical Effort

If we want to be more precise in our sizing, we can follow the rules outlined in Ref. [27]. Precharge circuit sizing is the distribution of electrical effort across sequential precharge leaf cells. We define a leaf cell here as the dynamic pulldown stack and the output inverter. It has been shown (see the Appendix) that the minimum delay for a series of precharge gates occurs when the stage effort for each gate is the same. So using logical effort to size our gates, we can follow this procedure:

1. Compute the path effort, F.
2. Estimate the best number of stages, N.
3. Estimate the least delay, D.
4. Determine the best stage effort, f_{min}.
5. Find gate sizes, C_{in}.
6. If the timing is not satisfactory, change the logical organization of the gates and repeat the procedure.

3.4   Sizing of the Keeper Device

As was discussed in Section 1.7, in the evaluation phase, if none of the inputs activate the NFET stack, the dynamic node must be kept HIGH. We use a keeper (weak PFET) to maintain the HIGH logic level. The purpose of the keeper is to fight leakage, not to fight any larger noise effects such as charge sharing.

Not only must a dynamic circuit be kept at a proper HIGH voltage level, it must also be writeable [36]. A keeper device thus must not be too weak and must not be too strong. These two criteria lead to a limitation as to how many pulldown stacks a circuit may have. Examples are wide OR gates and read bitlines in full-swing memory structures.

See Figure 3.1 for a better understanding.

Considering a 1 V supply, typical criteria established for the analysis are as follows:

1. The dynamic node must not droop **5%** below V_{dd} due to leakage (in the presence of 50 mV ground bounce). Corner = slow:fast (SF) (PMOS:NMOS).
2. The dynamic node must discharge from V_{in} at 50% to **5%** V_{dd} in **cycle/2** (in the presence of $V_{dd} - 50$ mV on the input). Corner = fast:slow (FS) (PMOS:NMOS).

One would run simulations for a combination of threshold voltages (high threshold voltage (HVT), standard threshold voltage (SVT), low threshold voltage (LVT)) and

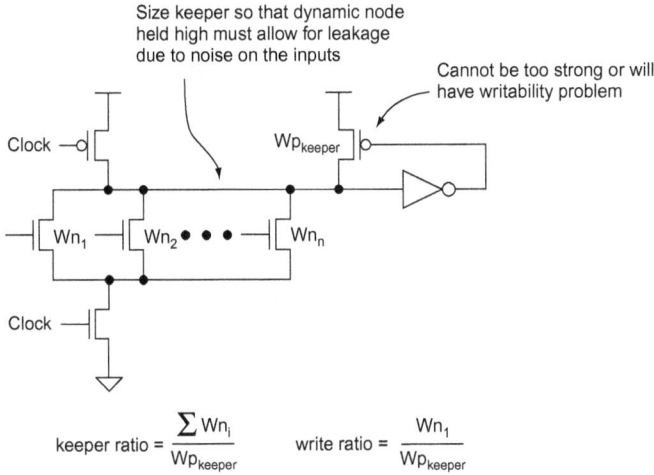

Size keeper so that dynamic node
held high must allow for leakage
due to noise on the inputs

Cannot be too strong or will
have writability problem

Clock

Wp_{keeper}

Wn_1 Wn_2 • • • Wn_n

Clock

$$\text{keeper ratio} = \frac{\sum Wn_i}{Wp_{keeper}} \qquad \text{write ratio} = \frac{Wn_1}{Wp_{keeper}}$$

Figure 3.1 Keeper ratio diagram.

different channel length devices. It is assumed that the keeper and pulldown stack have the same channel length. This is to minimize the PFET to NEFT mismatch, which is already "bad" because the diffusion shape, gate shape, gate material, and gate oxide are all different between NFETs and PFETs in modern processes. In any case, one should run simulations and fill in the tables like the examples shown below.

3.4.1 PFET Keeper

From the leakage and writability analysis, the following rules are derived for the PFET keeper: The minimum PFET keeper size is a percentage of the maximum leakable NFET width (see Section 3.4.2 on how to calculate this). Use appropriate percentages based on Table 3.1. In the event there is a mixture of different V_t or different L devices in the pulldown stack, use the data for the device that is OFF (leaking).

The maximum PFET keeper size is a percentage of the weakest pulldown path. Use appropriate percentages based on Table 3.2. In the event of a mixture of different V_t devices, use the largest V_t column. In the event of a mixture of different L devices, use the largest L.

Tables 3.1 and 3.2 give a range for the PFET keeper size. It is best to design for the smallest possible keeper so as not to sacrifice too much speed. If it turns out

Table 3.1 Minimum PFET Keeper Size (leakage)

Channel Length (nm)	HVT (%)	SVT (%)	LVT (%)
30	1	2	3
35	1	2	3

Table 3.2 Maximum PFET Keeper Size (writability)

Channel Length (nm)	HVT (%)	SVT (%)	LVT (%)
30	44	33	25
35	54	51	48

that the device size is out of range, then there are too many pulldown stacks on the dynamic node. Consider a different circuit topology and/or use DeMorgan's law to come up with different logic gates.

3.4.2   NFET Keeper

As discussed in Section 1.7, it is sometimes necessary to include an NFET keeper in the precharge gate. Considering the NFET keeper, here are typical rules: The minimum NFET keeper size is 1.5% of the precharge PFET width or one-fourth of the PFET keeper size—whichever is more. The maximum NFET keeper size is 15% of the precharge PFET width or equal to the PFET keeper—whichever is less.

3.4.3   Maximum Leakable NFET Width

The maximum leakable NFET width for each stack is determined based on two criteria:

1. Which device in the stack is the worst leaker?
2. What relationship exists between the stacks (mutual exclusivity)?

A simple technique for determining the leakable NFET width is as follows:

1. Pick the largest device in the stack.
2. Assume all other devices are ON.
3. Add the widths of the largest devices and use the total as the leakable NFET width.

A detailed example on leakable NFET width calculation:

In Figure 3.2, we see that the combination of $A = 0$ and $B = 1$ will cause node n2 to be fully discharged. It will also cause a $V_{ds} = V_{dd}$ to exist between node n1 and n2. The leakage for this input combination is worse than that for $A = 1$ and $B = 0$ due to the resistance of transistor A. However, simulations show that in a case such as the right evaluation stack, the input combination of $C = 1$, $D = 0$, and $E = 0$ has worse leakage than $C = 0$, $D = 1$, and $E = 1$.

Therefore, a simple approach for calculating the maximum leakable NFET width is to determine the maximum transistor width that can leak in each evaluation stack. For Figure 3.3, this would be 8 μm for the left evaluation stack and 12 μm for the right evaluation stack, for a total of 20 μm. For the left evaluation stack, this is a moderate overestimation of leakage current as the resistivity of device A is ignored, but the error should be small.

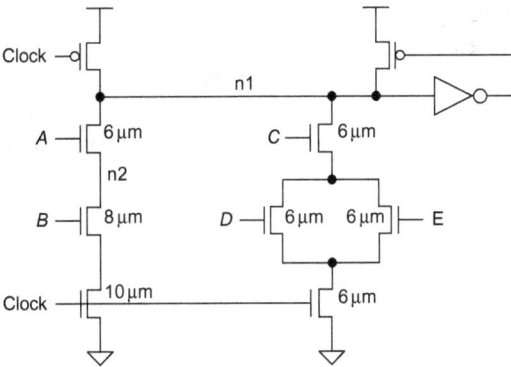

Figure 3.2 Maximum leakable N-width calculation.

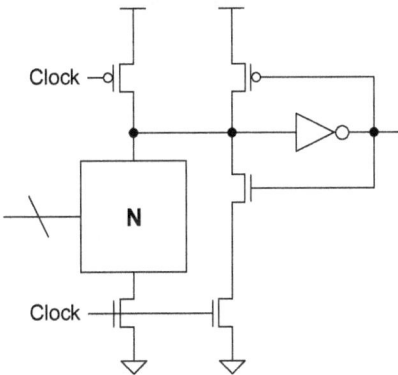

Figure 3.3 Clock-gated NFET keeper.

3.5   Sizing of the Precharge Device

Sizing of the precharge device depends on various timing considerations, including self-time margins for crowbar current and delay pushout (see Section 2.7). One can relax the precharge timing as long as one meets all edge rate and timing margin requirements. The precharge delay should be measured from the 50% point of the clock falling to 10% of the falling inverter output. The precharge device must be sized such that the rising edge rate is no more than approximately half of a phase.

If one uses a full keeper (PMOS and NMOS), the precharge edge can be too slow. In that event, try gating the NFET keeper with an evaluation device, as shown in Figure 3.3.

3.6   Sizing Precharge Gates with Wires

Sizing precharge gates in the presence of wires is more difficult than sizing for gate capacitance alone. This is because the resistance of the wires shields the capacitance on the far end of the wire, making it much more difficult to drive. It is possible that long wires can have a delay that is equivalent to a lumped capacitor two to six times that is predicted by the total capacitance of the wire. Because the RC of such long wires is determined by the output impedance of the driver, the wire characteristics, and the size of the receiving gate, a simple approach is needed.

Here is a simple procedure:

1. Obtain the total wire capacitance.
2. Convert the wire capacitance to gate load (~ 1 ff/1 μm in modern processes).
3. Add any diffusion and gate capacitance that is on the wire.
4. Size the driver for a fanout between 6 and 8.
5. Run a simulation to make sure that there is no violation of the minimum edge rate at the driver and the maximum edge rate at the receiver.
6. Refine the sizing for better delays. It is best to do a sweep of device sizes.

4 Noise Tolerance

Precharge logic gates have a much lower noise margin than conventional static CMOS. The noise margin is the amount of noise that can be introduced between a set of logic gates without causing the output gate to switch logic states. Static gates have relatively large noise margins ($NM_h = V_{oh} - V_{ih}$ and $NM_l = V_{il} - V_{ol}$). The noise can even exceed the threshold of a transistor without causing the logic gate to switch. In precharge logic, the situation is different. Because the dynamic node is not always driven, any noise that exceeds the threshold of an NMOS device can turn that device on and create a path from the dynamic node to V_{ss}, discharging the node. The evaluation part of the transfer characteristic transitions at a small input voltage is shown in Figure 4.1B.

The actual input noise limits depend on the capacitive load of the gate and also input noise pulsewidth. For example, a larger peak coupling noise could be permitted if the peak coupling occurred for only a fraction of a gate delay [8]. In a precharge gate, we need to consider the noise on the input to the gate as well as the noise on the dynamic node. Noise on the dynamic node can be made up of several components, including ground bounce, leakage (voltage droop), charge sharing, capacitive coupling, as well as noise propagated from the input.

4.1   Input-Connected Prechargers

Using input-connected prechargers can help noise tolerance on dynamic circuit inputs. An example of this technique is shown in Figure 4.2. The circuit shown below assumes that B does not reset until the end of the evaluation phase. Otherwise, the gate would reset prematurely.

Overall, these devices have more disadvantages than advantages:

Advantages:
- Noise limits on the inputs are higher.
- Noise on the dynamic net is reduced.

Disadvantages:
- Loading on input signals is increased.
- Layout requires more area for the extra devices and routing.
- There is more parasitic capacitance on the NMOS stack.
- Tricky timing when it comes to precharge.
- Sneak path problems exist in parallel evaluation stacks.

4.2   Propagated Noise

Noise can be a problem if it travels through multiple stages of logic gates. Transistors are good amplifiers and unfortunately, they also amplify noise. This can be seen clearly in Figure 4.3, where the NMOS device in the middle amplifies noise present on its input and passes the larger signal to the next gate. The noise can get big enough to discharge a dynamic node or cause a gate to switch to a wrong logic state.

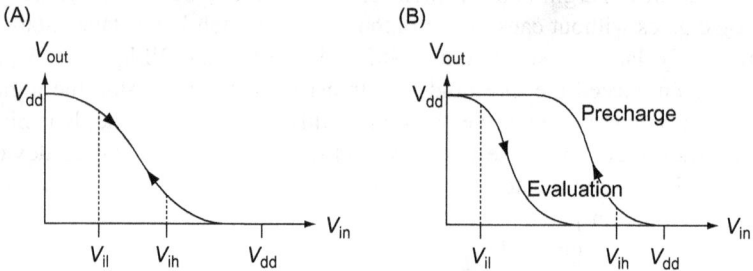

Figure 4.1 Transfer characteristics of (A) static and (B) precharge logic.

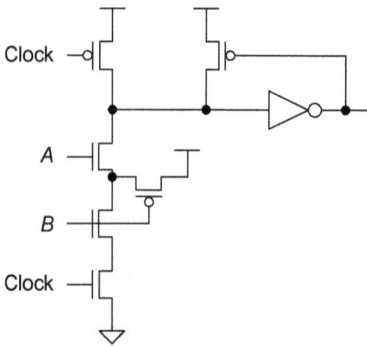

Figure 4.2 Input-connected precharger.

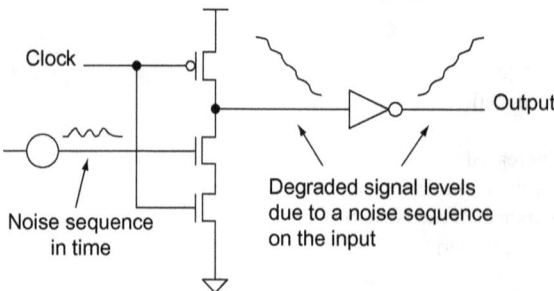

Figure 4.3 Noise amplification.

4.3   Input Wire Noise

There are two types of wire noise:

- Crosstalk (capacitive coupling) between an input wire and other wires, resulting in a positive glitch. A positive glitch can turn on an NMOS device and inadvertently discharge the
  dynamic node. This is shown in Figure 4.4A.
- IR drops on input wires degrade the signal level, which may cause slower transistor
  switching and may degrade the noise margins. This is shown in Figure 4.4B.

Capacitive coupling is an increasing problem as the height to separation ratio of
wires increases in modern submicron technologies. This increases the coupling
capacitance to other capacitances. For example, even in a 0.8 μm process, over
50% of the capacitance of a minimum pitch wire is to adjacent wires.

Problems with noisy wires usually are minimized by keeping input wires short
such that they do not experience too much capacitive coupling or too high a voltage
drop. IR drop also can be alleviated by making the wire wider, while capacitive
coupling problems can be mitigated by shielding the wire with power lines on both
sides. If a logic gate is fed directly by a dynamic latch, then the wire between them
must also be short so as not to disturb any charge stored on that node (wire). If a
wire cannot be shortened by layout constraints, then repeaters should be placed
along the path [33]. The repeaters restore a signal to proper levels, eliminating
much of the noise. The repeaters also reduce the propagation delay of a long wire.
If we have a wire with a lumped resistance and capacitance RC, then one repeater
results in two segments with $R/2$ and $C/2$. The total delay is thus half of what it
would be without the repeater, as shown in Eq. (4.1). The gate delay of the repeaters must be included in any delay calculations. Proper repeater placement involves
considerations of wire length, pitch, resistance, capacitance, load capacitance, and
driver output resistance.

$$2\left(\frac{R}{2} \cdot \frac{C}{2}\right) = \frac{RC}{2} \qquad\qquad (4.1)$$

Capacitive coupling results from the existence of mutual capacitances between
neighboring wires. Inductive coupling is presumed to be negligible in signal wires.

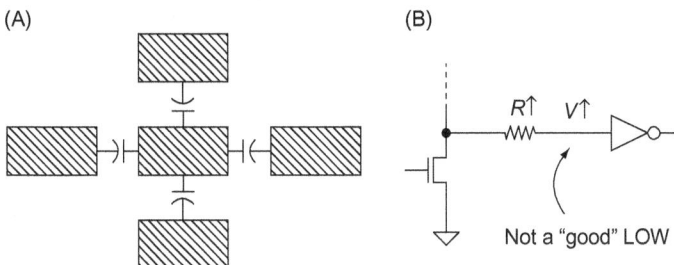

Figure 4.4 Capacitive coupling (A) and IR drops (B) in interconnect.

Capacitive coupling, on the other hand, increases with decreasing distance between wires, with increasing distance the wires are from the substrate, with increasing the length of neighborship, and with increasing frequency or edge rate of the signals [7].

In Ref. [8], there is an equation for the maximum length during which a wire can have a neighbor on a side. That equation is repeated here as Eq. (4.2). Where L_x is the length that the wires are neighbors, C_g is the total of *gate/drain/source* capacitances on the wire in question, C_s is the capacitance from the wire to the substrate, C_c is the effective coupling capacitance between the wires Eq. (4.3), C_m is the mutual capacitance between the wires, and L is the length of the wire in question. All capacitances can be per unit length. "s" is the switching factor and it is set to 1 if only one of the wires can switch and 2 if both wires can switch. The worst case is always assumed, where both wires switch in opposite directions. So the worst-case capacitance between the wires doubles if they both can switch. If the wire in question has a neighbor on each side and both neighbors switch in the same direction and opposite the wire in question, "s" is *not* set to 3. The reason is that the extra neighbor is already taken into account by C_m.

L_x can have a maximum value of $2L$ in which case there are two neighbors running the full length of the wire (one on each side). This implies that it is sometimes better to lengthen a wire to lower the noise. This is true as long as the extra length does not receive any coupling and only sees constant potentials such as the substrate. In addition, routing of unrelated signals should be minimized.

$$L_x \leq \frac{(4C_g/C_s) + L}{(\Delta V_{dd}/V_{dd}) \cdot (C_c/C_s)} \tag{4.2}$$

$$C_c = sC_m = s\frac{A\varepsilon}{d} \tag{4.3}$$

Spacing between wires can be increased if coupling is too high. Minimum wire spacing can be determined by placing an upper bound on the coupling between wires and knowing the dielectric constant between them. Equation (4.4) [7] can be used to determine minimum wire spacing using the coupling capacitance and is valid if the wires have a homogenous dielectric between them. K_v is the amount of voltage coupling, V_q is the induced voltage on the wire of interest, and V is the change in voltage on a neighboring wire. C_m and C_c are as defined above and C is the capacitance between a wire and the rest of the system (assume the same for both wires).

$$K_v = \frac{V_q}{V} = \frac{C_c}{C} \tag{4.4}$$

Another common solution to capacitive coupling on wires, especially in datapaths, is to twist the lines as shown in Figure 4.5 [2]. This allows us to take advantage of power rails and reduce the noise on multiple wires with one power line. We would route dynamic lines between mutually exclusive and complementary lines.

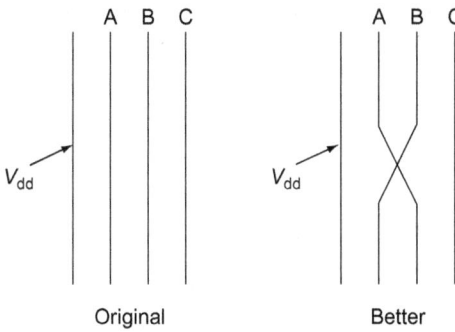

Figure 4.5 Twisted wires to reduce noise.

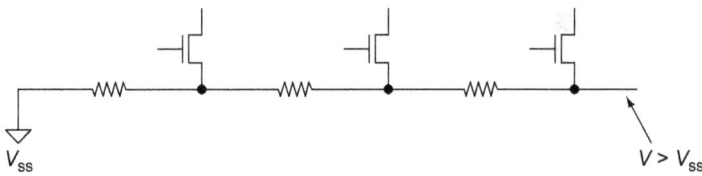

Figure 4.6 Ground bounce.

Finally, in Ref. [34], the authors discuss a technique for recoding busses as one hot for fewer transitions. This reduces the amount of capacitive coupling. As a final word on wire noise, another solution is to use high-threshold NMOS devices in the receiver.

4.4 Supply-Level Variations

There are two kinds of supply-level variations:

- Power rail IR drop between driver and receiver (Figure 4.6).
- External noise on the supply of the chip (Figure 4.6).

Modern chips have trends toward lower voltage and higher power. Higher power increases the supply current. Lower voltage also increases the supply current if the power is held constant. Therefore, the current is increasing quadratically. For a fixed current, the *IR* drop in a power supply wire increases as a fraction of the supply voltage. So power supply noise is getting worse cubically if the resistance in the power rail is constant.

The largest *IR* drop in the power rails comes from many gates switching at the same time (especially when the clock switches). This is known as simultaneous switching noise, also known as ΔI noise or ground bounce. Current transients can generate large potential drops in the power rails due to the inductance and resistance present in the power rails. The inductive drop comes from Eq. (4.5), also known as Faraday's law. Remedies include limitations on how many large devices switch at the same time (in the same region) and on decoupling capacitance [8].

The resistive drop can be lowered by using stiffer power rails (lower impedance). All power supply variation effects can be minimized by placing gates closer together, i.e., avoiding long signal wires.

$$V = L\frac{dI}{dt} \tag{4.5}$$

Supply noise can be specified in terms of the inductance in the power rails. This can be accurate, but may be too difficult to do for a large system. Equation (4.6) can be used to determine the maximum amount of noise on V_{ss} [8]. B is the number of gates that are switching, L is the inductance in V_{ss}, the capacitances are shown in Figure 4.7A, $t_{r,in}$ is the rise time of the input signal, t is a linear approximation of the rise time at 50% of V_{dd}, and $\beta_n = \mu C_{ox}$.

$$\Delta V_{ss}(max) = x = \frac{BLC_p\beta_n}{(C_p + C_n)t_{r,in}}\left[\frac{t}{t_{r,in}}(V_{dd} - x) - V_t - x\right](V_{dd} - x) \tag{4.6}$$

Decoupling capacitors decrease the noise on power rails and are used only if supply noise is very high. Figure 4.7A shows how a decoupling capacitor is inserted into the system. The relatively large decoupling capacitor acts like a power supply during circuit switching. It supplies or removes any additional current and thus stabilizes the supply voltage. The capacitor leads and other dimensions must be small to minimize parasitic inductances [7]. The value of the capacitor has to be large enough to limit supply fluctuations to a desired range, yet small enough not to lower the system resonant frequency close to the operating frequency [7].

Equation (4.7) can be used to determine an appropriate value for the decoupling capacitance given maximum allowable supply fluctuations, ΔV_{dd}. Equation (4.7) is an approximation because it does not take any inductance into account. The inductance is assumed to be infinite so that no current flows through the inductor (power rails) immediately after switching. Thus, the circuit reduces to a capacitive divider, as shown in Figure 4.7A.

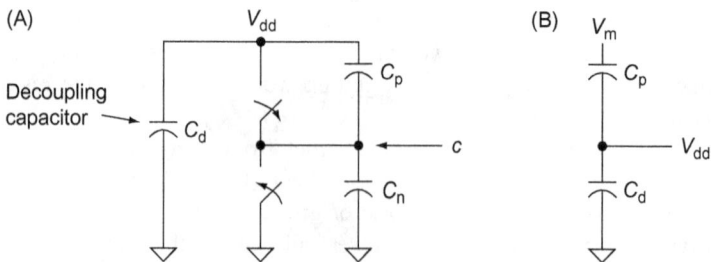

Figure 4.7 System with a decoupling capacitor (A), and redrawn circuit for capacitor calculation (B).

The easiest way to estimate a value for the decoupling capacitor is to redraw the circuit, as shown in Figure 4.7B. The capacitors can be viewed as being in series because V_{dd} does not behave like a supply but rather as a node between two capacitors as the instant switching takes place. Voltage V_m transitions from V_{dd} to ground, which leads to Eq. (4.8). This equation can be used easily to estimate a value for the decoupling capacitor. When doing an actual calculation, it is important to model accurately all the capacitances correctly.

$$\Delta V_{dd} = \Delta V_m \frac{C_p}{C_p + C_d} \tag{4.7}$$

$$\Delta V_{dd} = -V_{dd} \frac{C_p}{C_p + C_d} \tag{4.8}$$

4.5   Charge Sharing

Charge sharing is a problem that exists in precharge logic gates that contain series stacks of transistors [6]. This is best explained with an example in Figure 4.8. If A and B go HIGH but C remains LOW (during the evaluate phase), some charge from C_{out} will redistribute into C_1 and C_2. So the dynamic node X loses some of its charge, and this is what is meant by charge sharing. If node X loses too much charge, then it could be interpreted as a logic LOW rather than a HIGH, and the static inverter could make an unwanted transition.

Charge sharing is a problem for dynamic circuits because of their inability to recover from glitches. The charge loss can be very significant even when the dynamic node charge shares to just one intermediate node. Only a small percentage of V_{dd} typically is allowed for charge sharing on dynamic nodes, including bitlines

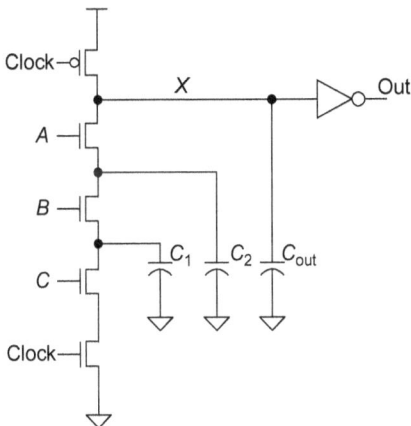

Figure 4.8 Capacitances in a precharge gate.

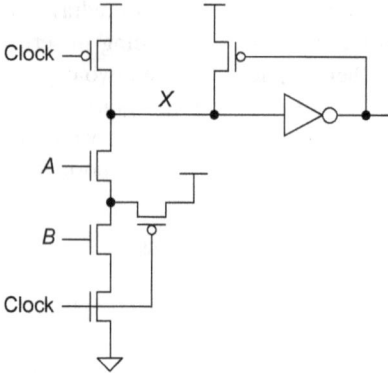

Figure 4.9 Internal node precharger.

and matchlines in full-swing RAMs. Charge sharing is only a problem if C_{out} is *not* much greater than $C_1 + C_2$. To prevent charge sharing, extra precharge devices can be added to precharge some of the intermediate nodes, as shown in Figure 4.9. If the intermediate nodes are precharged just like node X, then they have no reason to take charge from node X. Precharging every other node is generally sufficient to keep charge sharing to about 10% of V_{dd} ().

So if the dynamic net capacitance is too small, one will need internal node precharge devices to prevent charge sharing (Figure 4.8). There are some advantages and disadvantages in using these devices:

Advantages:
- Charge sharing is prevented.
- Precharge timing is more predictable.
Disadvantages:
- Layout requires more area for the extra devices and routing.
- There is more parasitic capacitance on the NMOS stack.

Finally, it should be noted that not all nodes pose a problem in terms of charge sharing (Figure 4.10).

4.6   Charge Sharing: Example 1

An example of charge sharing will now be presented. The circuit and timing for this example are shown in Figures 4.11 and 4.12. This example considers what happens when input B transitions rapidly from LOW to HIGH, input A remaining LOW. As will be shown, this situation can lead to charge sharing.

The calculations must be done in two parts. First, there will be crosstalk through transistor B, resulting in some voltage changes on node X and the output, just after input B goes HIGH. Then transistor B will turn on and all charges will attempt to equalize. From the circuit diagram, we can write Eq. (4.9), which shows the basic

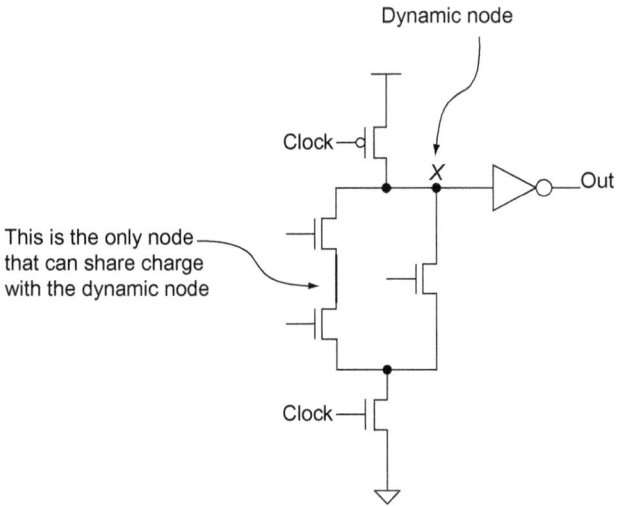

Figure 4.10 Not all nodes are a problem for charge sharing.

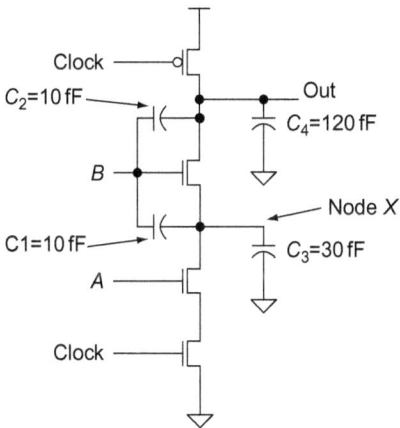

Figure 4.11 Circuit for the example of charge sharing.

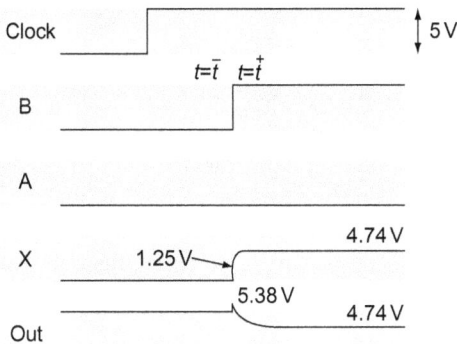

Figure 4.12 Timing for the example of charge sharing.

relationship between voltages V_B and V_X: charges stored in two capacitors in series are equal. After some manipulation, we end up with Eq. (4.10).

$$C_1 \Delta(V_B - V_X) = C_3 \Delta V_X \tag{4.9}$$

$$C_1 \Delta V_B - C_1 \Delta V_X = C_3 \Delta V_X$$

$$\frac{C_1}{C_1 + C_2} \Delta V_B = \Delta V_X \tag{4.10}$$

$$\therefore \frac{10 \text{ fF}}{10 \text{ fF} + 30 \text{ fF}} (5 \text{ V}) = 1.25 \text{ V}$$

So the change in V_X due to a 5 V change in V_B is 1.25 V. And now to calculate the final value of node V_X, we write Eq. (4.11). Node X is originally at 0 V, so the final voltage on node X is 1.25 V.

$$V_X(t^+) = V_X(t^-) + \Delta V_X \tag{4.11}$$

$$\therefore V_X(t^+) = 0 \text{ V} + 1.25 \text{ V} = 1.25 \text{ V}$$

Similarly, for the output node, we can write Eq. (4.12). The resulting change in voltage is 0.38 V. The output was precharged to 5 V before input B made its transition. The final value for the output is calculated using Eq. (4.13), and the result is 5.38 V.

$$\frac{C_2}{C_2 + C_4} \Delta V_B = \Delta V_{\text{OUT}} \tag{4.12}$$

$$\therefore \frac{10 \text{ fF}}{10 \text{ fF} + 120 \text{ fF}} (5 \text{ V}) = 0.38 \text{ V}$$

$$V_{\text{OUT}}(t^+) = V_{\text{OUT}}(t^-) + \Delta V_{\text{OUT}} \tag{4.13}$$

$$\therefore V_{\text{OUT}}(t^+) = 5 \text{ V} + 0.38 \text{ V} = 5.38 \text{ V}$$

The second part of the calculation focuses on final voltage levels after a long time has passed. Once transistor B turns on, charge redistribution between the output node and node X takes place. But we must remember that node X cannot rise above $V_{\text{OUT}} - V_{\text{tn}}$, so there is a limit as to how much the charges will redistribute. It is best to start with the assumption that the two nodes are connected (through M_B) and the charges can redistribute without any constraints. In this case, the only way to analyze the system is to add the charges before and after M_B switches, as shown in Eq. (4.14). Calculations are first done for $t = t^+$, i.e., right after input B switches, as shown in Eq. (4.15).

$$\sum_{n=0}^{m} (V_C(t+) \cdot C)_n = \sum_{n=0}^{m} (V_C(\infty) \cdot C)_n \tag{4.14}$$

$$t = t^+: \quad Q(t^+) = (V_B - V_{OUT})C_2 + V_{OUT}C_4 + (V_B - V_X)C_1 + V_X C_3$$
$$= (5 - 5.38)10 \text{ fF} + 5.38 \times 120 \text{ fF} + (5 - 1.25)10 \text{ fF} + 1.25 \times 30 \text{ fF} \tag{4.15}$$

$$\therefore Q(t^+) = 716.8 \text{ fC}$$

Next, calculations will be done for $t = \infty$, and then the charges will be set equal. We start by writing Eq. (4.16) which is a modification of Eq. (4.15) with each V_{OUT} and V_X replaced by V_{NEW}.

$$t = \infty: \quad Q(\infty) = (V_B - V_{NEW})C_2 + V_{NEW}C_4 + (V_B - V_{NEW})C_1 + V_{NEW}C_3$$
$$= (5 - V_{NEW})10 \text{ fF} + V_{NEW} \times 120 \text{ fF} + (5 - V_{NEW})10 \text{ fF} + V_{NEW} \times 30 \text{ fF} \tag{4.16}$$

Because

$$Q(\infty) = Q(t^+)$$

$$\therefore V_{NEW} = 4.74 \text{ V} = V_X(\infty) = V_{OUT}(\infty)$$

4.7 Charge Sharing: Example 2

In this example, we will assume that the inputs make a gradual transition so that there is no transistor crosstalk. We will further assume that the gate to drain/source capacitances are negligible to keep the calculations simple. We will, however, consider what happens when only a small amount of charge is shared between the output node and the intermediate nodes. As we will see, in this case the voltages do not fully equalize. The circuit for this example is shown in Figure 4.13.

The above figure shows the charge-sharing circuit that occurs during evaluation. We can see that C_{OUT} must share charge with C_1 and C_2. The final voltage across these capacitors will be V_{NEW}. Two possibilities exist, depending on the level of charge sharing. If $V_{NEW} < V_{dd} - V_{tn}$, then all the capacitors will have the same voltage and the final charge distribution is given in Eq. (4.17), where V_{OUT} is V_{dd}.

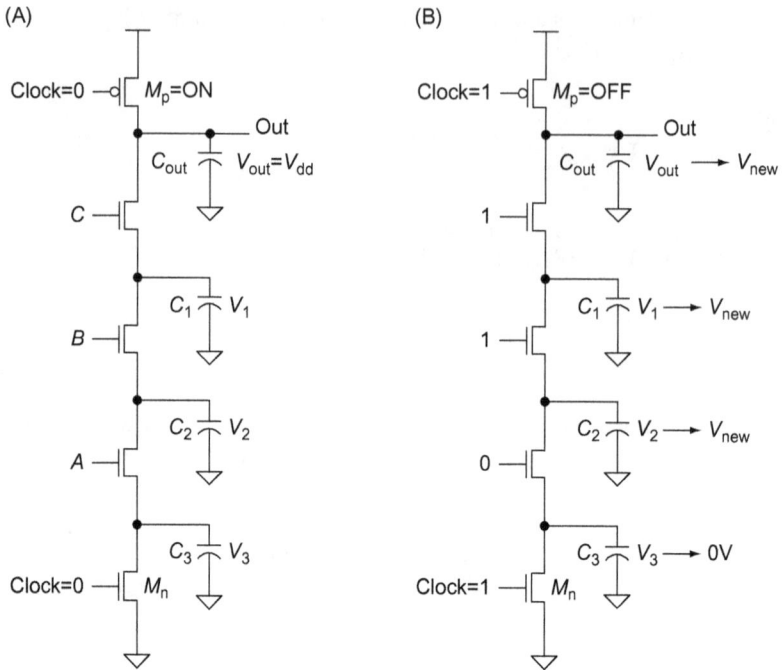

Figure 4.13 Circuit for the example of charge sharing, precharge network (A), and charge-sharing circuit (B).

$$Q = C_{OUT}V_{NEW} + C_1 V_{NEW} + C_2 V_{NEW} = C_{OUT}V_{OUT} \qquad (4.17)$$

The final voltage is thus as given in Eq. (4.18).

$$V_{NEW} = \frac{C_{OUT}}{C_{OUT} + C_1 + C_2} V_{dd} \qquad (4.18)$$

So if we want to have a logic 1 value on the output, we must have $C_{OUT} \gg C_1 + C_2$. Thus, care must be taken in layout to keep C_1 and C_2 small.

The next possibility is that only a small amount of charge is lost from C_{OUT} and $V_{NEW} \geq V_{dd} - V_{tn}$. Because M_1 and M_2 are effectively pass transistors, the maximum voltage that can be transferred to C_1 and C_2 is $V_{dd} - V_{tn}$. So the final charge distribution is as follows in Eq. (4.19):

$$Q = C_{OUT}V_{NEW} + (C_1 + C_2)(V_{dd} - V_{tn}) = C_{OUT}V_{OUT} \qquad (4.19)$$

So V_{NEW} *on the output node* is then as given in Eq. (4.20).

$$V_{NEW} = V_{dd} - \frac{C_1 + C_2}{C_{OUT}}(V_{dd} - V_{tn}) \qquad (4.20)$$

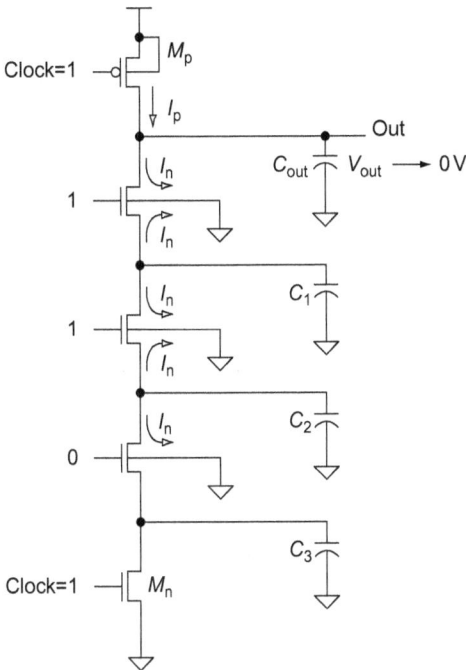

Figure 4.14 Leakage paths.

4.8 Leakage

Leakage comes from several sources, with the subthreshold leakage being the largest. All are modeled in modern transistor models from foundries:

- Subthreshold leakage in the evaluation stack.
- Diode leakage on the dynamic node.
- Gate leakage from the output inverter.

Normally keepers are used to fight the leakage and maintain the HIGH voltage level on the dynamic node. Keeper design was discussed in Sections 1.7 and 3.4. We shall now take a look at a leakage calculation example. Voltage droop due to leakage can be checked with Eq. (4.21).

$$V_{\text{droop}} = \frac{I_{\text{leak}} \cdot t}{C_{\text{tot}}} \tag{4.21}$$

The circuit for this example is shown in Figure 4.14.

The above figure shows the worst-case leakage paths that occur during evaluation. Every pn junction has reverse current flow. M_p provides I_p to charge C_{OUT}. However, three NFETs drain charge to ground. This gives Eq. (4.22) for the discharge of C_{OUT}.

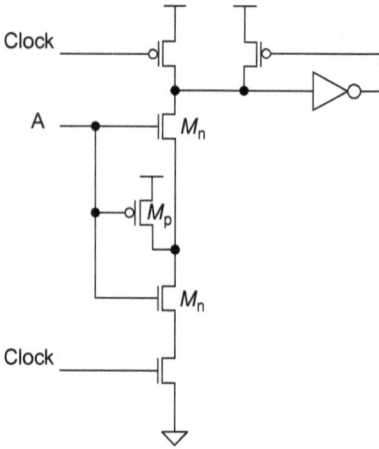

Figure 4.15 The stack effect to reduce leakage.

$$5I_n - I_p = -\ (C_{OUT} + C_1 + C_2)\frac{dV_{OUT}}{dt} \tag{4.22}$$

The assumption is that all the capacitors are at the same voltage. The time to discharge C_{OUT} completely is thus given by Eq. (4.23).

$$dt = 5I_n - I_p + (C_{OUT} + C_1 + C_2)V_{dd} \tag{4.23}$$

In addition to keepers, several other things can be done to reduce the leakage in a precharge gate [36]:

• Increased channel length (reduces channel leakage, increases gate oxide leakage)
• Stack effect
• High V_t transistors
• Back bias
• Reducing the beta ratio of the output stage

An example of the "stack effect" [46] is shown in Figure 4.15. When the input is LOW, M_p biases the stack node so that M_n has negative V_{gs}. This shuts off the leakage current.

4.9 Clock Coupling on the Internal Dynamic Node

When the clock goes LOW, the internal dynamic node can get coupled below V_{ss}, as shown in Figure 4.16. This creates substrate current problems and has precharge timing implications. This problem can be solved by precharging the virtual ground node to V_{dd}, using an internal node precharge device.

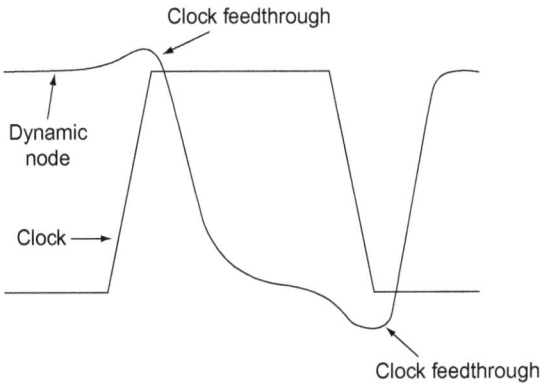

Clock feedthrough

Figure 4.16 Clock feedthrough.

Dynamic node

Clock

Clock feedthrough

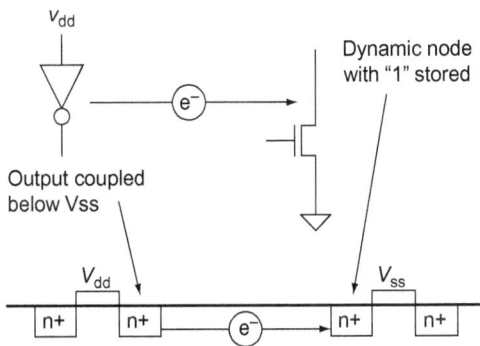

Figure 4.17 Minority carrier charge injection.

Dynamic node with "1" stored

Output coupled below Vss

4.10   Minority Carrier Charge Injection

Minority carrier charge injection can happen when a logic gate output is coupled below V_{ss} [2]. In such a case, the drain–substrate junction becomes forward biased and injects charge into the substrate. The problem is when that driver is close to a dynamic node, as shown in Figures 4.17 and 4.18. The fix for such problems is to place a diffusion collector tied to V_{dd}. This kind of collector is sometimes referred to as a guard ring, if it surrounds the sensitive circuit entirely.

4.11   Alpha Particles

Alpha particles are from space. These are helium nuclei emitted from radioactive decay. When they crash into chips they leave a trail of electron-hole pairs in the substrate [35]. These carriers are collected by dynamic nodes, disturbing the

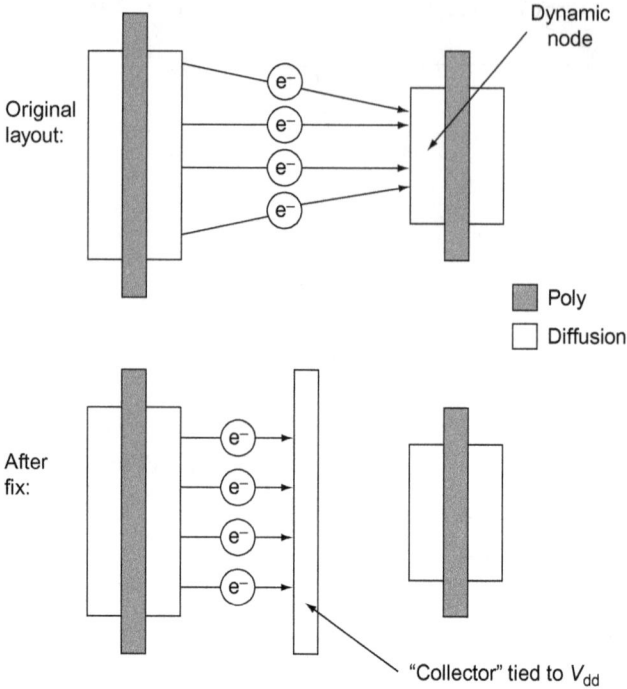

Figure 4.18 Charge injection layout fix.

voltage on the node. These are referred to as "soft errors". Alpha particle hits are only a problem for very small precharge gates and memories. Soft errors are an increasing problem as technology shrinks and the dynamic node capacitance gets smaller.

4.12 Noise Induced on Dynamic Nodes Directly

There are a number of noise sources that impact the dynamic node directly. Some of these have already been discussed in detail. Following is a summary of all of these noise sources:

- Propagated noise from the inputs, discussed in Section 4.2.
- Capacitive coupling [2], similar to the discussion in Section 4.3.
- Supply-level variations, discussed in Section 4.4.
- Charge sharing, discussed in Sections 4.5–4.7.
- Device leakage, discussed in Section 4.8.
- Electromagnetic radiation from external sources.
- Clock coupling, discussed in Section 4.9.
- Minority carrier charge injection, discussed in Section 4.10.
- Alpha particles from space, discussed in Section 4.11.

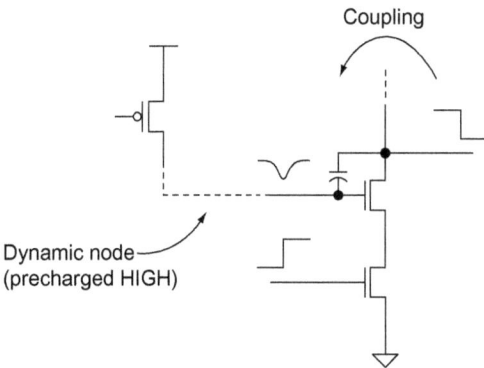

Figure 4.19 Output transition back coupling to a dynamic node.

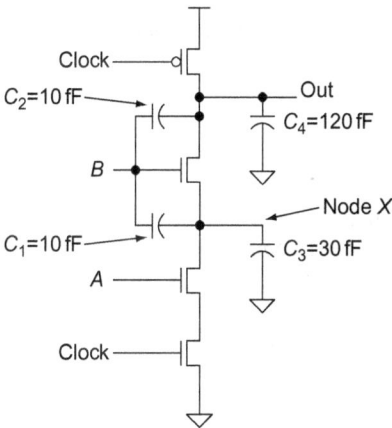

Figure 4.20 Circuit for the example of transistor crosstalk.

- Output transition back coupling, as shown in Figure 4.19.
- Input crosstalk forward coupling, as will be discussed here.
- Input HIGH-to-LOW transition forward coupling during precharge, as in Section 4.13.

Crosstalk on an input wire can result in a negative glitch. Such a negative glitch can be transmitted to a dynamic node through the gate to drain capacitance. This is referred to as input crosstalk forward coupling, and will be discussed in Section 4.13.

4.13   Example of Transistor Crosstalk During Precharge

This example deals with crosstalk through the gate to channel capacitances in a MOSFET. The circuit and timing for this example are shown in Figures 4.20 and 4.21. This example deals with changes in voltage levels when devices switch off during a precharge phase. It shows how capacitive coupling affects a precharged gate. Capacitive coupling simply means that if one side of a capacitor changes in potential very quickly, the other side will do the same. This is due to the fact that

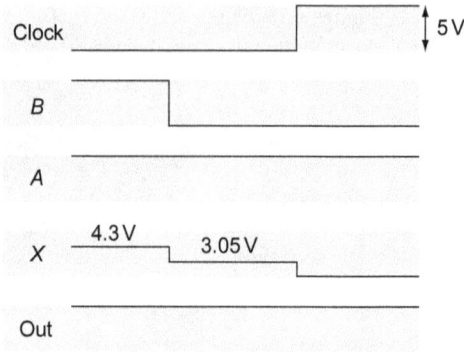

Figure 4.21 Timing for the example of transistor crosstalk.

the current through a capacitor is equal to the derivative of the voltage across it, as shown in Eq. (4.24). If the voltage across the capacitor changed instantly, then the current through it would have to be infinite, which is impossible.

$$I = C\frac{dV}{dt}$$
(4.24)

From the circuit diagram, we can write Eq. (4.25) because C_1 and C_3 are in series. Into Eq. (4.25), we can substitute some values to compute a result.

$$\frac{C_1}{C_1 + C_3}\Delta V_B = \Delta V_X$$
(4.25)

$$\therefore \frac{10 \text{ fF}}{10 \text{ fF} + 30 \text{ fF}}(-5) = -1.25 \text{ V}$$

So the change in V_X due to a -5 V change in V_B is -1.25 V. And now to calculate the final value of node X, we write Eq. (4.26). Node X is precharged to 4.3 V because we are assuming V_{tn} to be 0.7 V. So the final voltage on node X is 3.05 V.

$$V_X(t^+) = V_X(t^-) + \Delta V_X$$
(4.26)

$$\therefore V_X(t^+) = 4.3 \text{ V} - 1.25 \text{ V} = 3.05 \text{ V}$$

4.14   CSR Latch Signal Ordering

From a timing perspective, it is best to connect the dynamic net to the NANDx input that is closest to the output of that gate. However, because coupling from the dynamic net to the output of the NANDx is a concern, we will connect the dynamic net to the side that is closest to V_{ss}. This is specified with a dot in Figure 4.22.

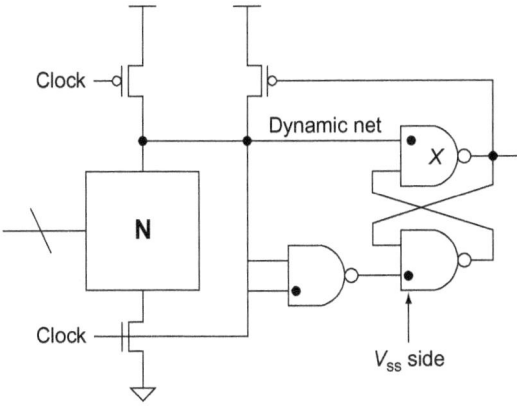

Figure 4.22 CSR latch.

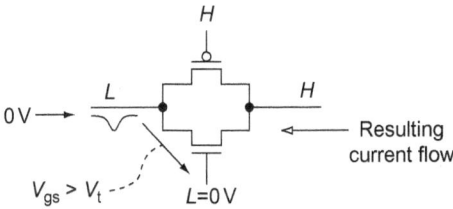

Figure 4.23 Dynamic latch discharge due to noise.

4.15   Interfacing to Transmission Gates

Transmission gates following precharge gates or on the inputs of precharge gates are generally not allowed in modern companies. All the noise sources discussed thus far can be particularly dangerous if the wire is feeding a dynamic transmission gate latch. This situation is shown in Figure 4.23, where the input to the latch is LOW and is taken below V_{ss}. The input node is the source of the NMOS device in the latch, and if it drops more than V_{tn} below V_{ss}, then the dynamic node on the other side will start to discharge. Furthermore, the junction diode in the device becomes forward biased and there is current injection into the substrate. This can lead to latchup.

Furthermore, simple charge sharing can occur when the transmission gate turns on. This can disrupt the dynamic node. This is another reason precharge gates should not drive the source/drain inputs of pass transistors or transmission gates.

5 Topology Considerations

5.1 Limitation on Device Stacking

In modern processes, it is inefficient to implement a function with more than four stacked devices in a single evaluation stack. This means one can implement an AND3 with an evaluation footer or an AND4 without the footer.

5.2 Limitation of Logic Width

The problem with wide precharge OR gates is that the keeper quickly becomes large with respect to the pulldowns. This creates writability problems. Additionally, the parasitic capacitance on the internal dynamic net will become significant (also a speed issue). So it is recommended that a wide dynamic OR gate be limited to no more than 8 inputs.

5.3 Use of Low/High V_t Transistors

Modern processes allow for three types of threshold voltages: low, standard, and high V_t. The inclusion of LVT devices in domino gates is a very good way to improve the evaluation timing. However, due to noise and leakage concerns, LVT devices can only be allowed in the following locations:

1. The evaluation footer.
2. The precharge device.
3. The output inverter.

Please note that using an LVT footer does not increase the leakage in a precharge gate. This is because the footer device is on during evaluations and the higher V_t logic devices in the stack determine the leakage current. HVT devices of course do not increase leakage, either.

We do not want to use LVT devices in the evaluation stack because their leakage current would increase considerably. As such, keeper sizing must be much larger to counter the effect of the large leakage current. Also, the noise immunity of the gate is decreased, as it is much easier to propagate noise through the dynamic gate.

Figure 5.1 shows the use of LVT devices in dynamic gates represented by circles around the LVT devices.

Synchronous Precharge Logic.
© 2012 Elsevier Inc. All rights reserved.

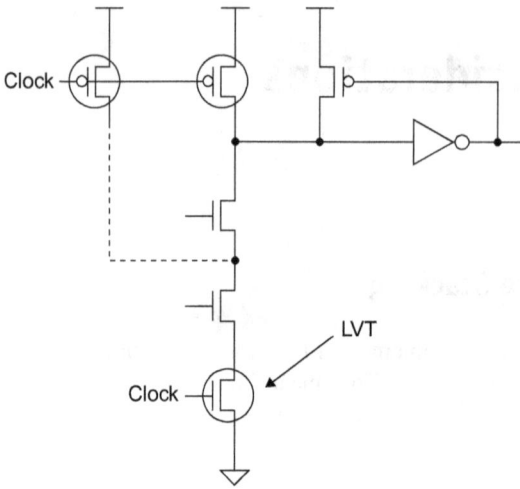

Figure 5.1 LVT usage.

5.4   Sharing Evaluation Devices

The use of a single shared evaluation footer may help or hurt performance, depending on the nature of the inputs. If the clock arrives first, the larger n2 capacitance creates a better virtual ground. The circuit will thus evaluate faster. If, on the other hand, the input arrives first, then n2 is not fully discharged before evaluation. The sharing of the evaluation device will hurt performance in this case because a larger n2 capacitance must be discharged before n1 is discharged (Figure 5.2).

Figure 5.3 shows a gate where sharing the evaluation device is not beneficial. It will slow the gate down because a similarly sized evaluation device will discharge more capacitance if it is shared than if it is not shared. A nonshared device will discharge either n2 or n3, while a shared device must discharge both.

If there is mutual exclusivity in the inputs to the previous gate, then sharing the footer might not slow down the gate due to the extra current from n2 and n3 discharging. In this case, a tapered footer could be used with good benefit for the area

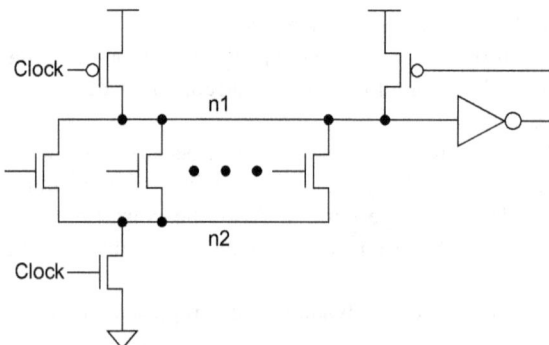

Figure 5.2 Sharing the footer device.

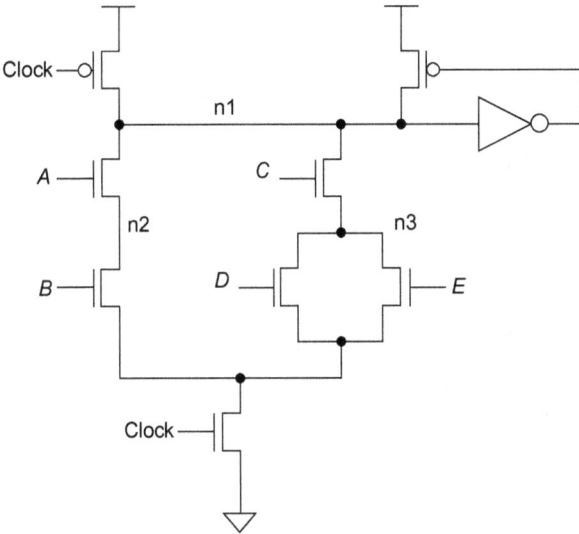

Figure 5.3 Not beneficial sharing of the footer device.

required. As a final note, for all gate configurations, the capacitance of the virtual ground node will be higher with shared footers than with nonshared.

5.5 Tapering of the Evaluation Device

Increasing the size of the evaluation footer (tapering) has some advantages and disadvantages.

Advantages:
- The conductivity of the stack is much better.
- The ability to hold the virtual ground node at V_{ss} improves significantly.

Disadvantages:
- The layout is more difficult to do (poly spacing, diffusion jog).
- The clock power for the gate increases.
- Timing requires that the clock be early, implying monotonic inputs.

In the extreme, a large tapering of the footer would provide the same benefit as removing the device entirely. However, we do not have the delay penalties associated with fast precharging, if we keep the footer.

5.6 Footed versus Unfooted

In a block of domino gates, each gate needs a precharge PMOS device [4]. But only the first gate needs an NMOS evaluation device, as shown in Figure 5.4. Signal a_r, that is, the output of the first domino gate is guaranteed LOW during

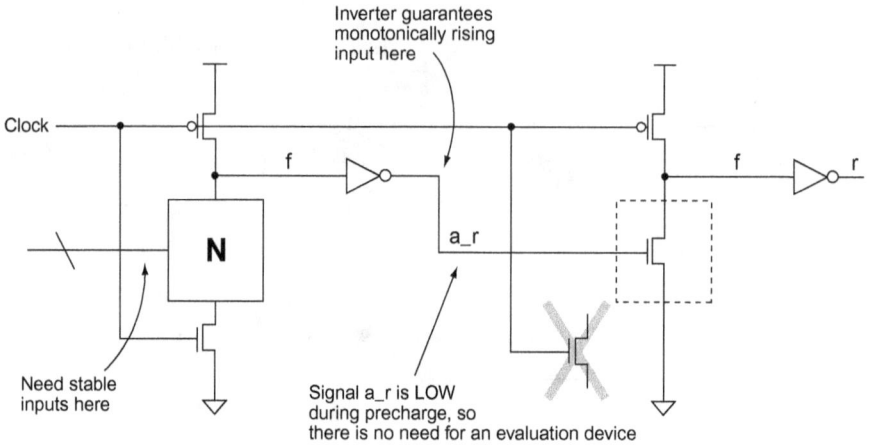

Figure 5.4 Only the first domino gate needs a footer.

precharge. So there is no need to include an evaluation footer with every logic gate. If it is not there, the gate is smaller and faster.

To illustrate the use of evaluation devices, let us have a look at Figure 5.5, which is a block of logic with multiple inputs. This diagram has two logic gates, one driving the other. The input transistor that receives this signal does not need an evaluation device, as explained above. The second gate also has an input from some random logic, which could be unclocked and arriving from a different clock domain. In this case, this input signal might not be monotonic in nature, and so an evaluation device is needed there.

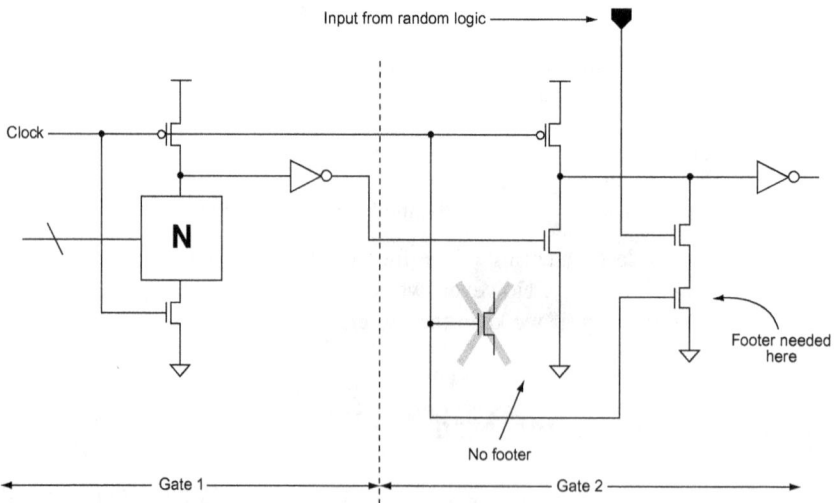

Figure 5.5 Use of an evaluation footer in a multiple input gate.

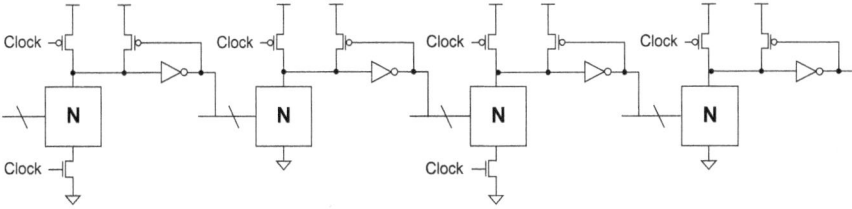

Figure 5.6 Footer sequence.

But what we have here is noninverting logic. If an external input rises, the output might also rise, but it will never fall due to that input. DeMorgan's laws can be used to push the inversions around [4]. Furthermore, it is possible to generate both the output and its complement with two precharged gates [4]. The point here is that if we want a certain output, that is active LOW, all we have to do is to solve the Boolean equation for that output being active HIGH. Implementation with NMOS devices provides the extra inversion. In general, the idea is to solve the Boolean equation for the complement of the signal wanted. As long as the laws of Boolean algebra are followed, we will have exactly the same logical expression.

So if one has a sequence of domino gates as shown in Figure 5.6, all the inputs to the gates (except the first one) are guaranteed to be monotonic. As mentioned above, one does not really need the footer evaluation device, which can speed up the evaluation. However, the problem is that there is extra power dissipation during precharge because there is some overlap time in when the inputs are active and precharge is already on. To avoid contention in an unfooted gate completely, the gate should not begin precharge until the previous gate is fully precharged. It is difficult to guarantee this across process corners. Therefore, it is generally recommended that the footer device be removed at most every other domino gate, as shown in Figure 5.6. We will call nonfooted precharge gates D2 gates.

Here are some more details on the advantages and disadvantages of removing the footer:

Advantages:
- Reduced stack height, which leads to area, speed, and power benefits.
- Reduced clock loading, which helps power and clock skew.

Disadvantages:
- Precharge timing requirements reduce the benefits significantly for two-phase clocking.
- Delayed clock elements must be designed to avoid short circuit current.
- Inputs to nonfooted gates must be monotonic.

By delaying the precharge clock to the D2 gate, it now has much less time to precharge. This can be seen in Figure 5.7. The precharge time is clearly reduced by t_d. This leads to a larger precharge device (speed penalty) and the output inverter will be limited in the fanout it can drive.

Although the speed benefit of removing the evaluation devices is obvious, it is offset by the secondary effects seen in the D2 gate. For this reason, removal of

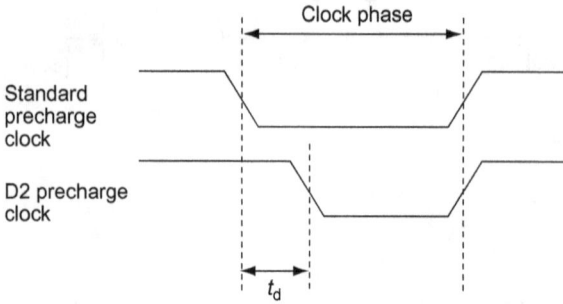

Figure 5.7 D2 precharge timing requirements.

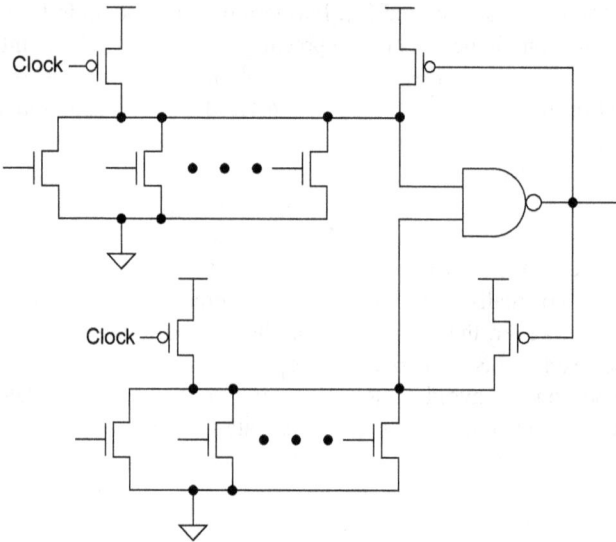

Figure 5.8 Compounded wide OR.

evaluation devices requires considerable design effort and review. Short circuit current must be avoided, as it will increase the power consumption of the chip and increase power supply variations (Figure 5.8).

5.7 Compounding Outputs

Precharge logic is better at NOR gates than NAND gates. This is because in a NOR gate, all transistors in the NMOS tree are in parallel, which minimizes the pulldown time. But what if one needs a NAND gate? Worse yet, what if it is a high fan-in NAND gate? The solution is to break up the NMOS tree and use a static NOR gate on the output instead of an inverter [4]. An example of a

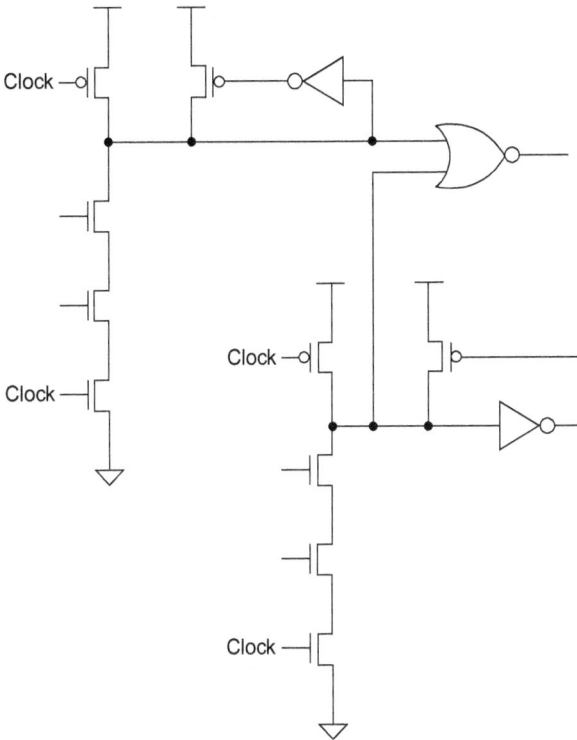

Figure 5.9 Compounded wide AND.

four-input AND gate is shown in Figure 5.9. This trick is useful even if implementing a high fan-in OR gate, because it reduces the drain capacitance on the dynamic node [4].

Compounding outputs have speed and Boolean benefits. For example, the use of a static NAND gate on the output of a dynamic OR gate allows for efficient construction of a wide OR function. One problem with the specific circuit in Figure 5.9 is that there is a floating node if one evaluation stack evaluates, while the other one does not. The problem is easily solved by using separate inverters for each keeper. Although this circuit change removes the floating node, it has a capacitance penalty. The floating node created in Figure 5.9 does not have power or functionality implications because the situation only occurs when the NAND has already transitioned to an output of logic one. As such, the floating node is not a problem in any way. Therefore, it is an allowable exception to the general preclusion against floating nodes.

Several factors need to be considered before a decision on compounding is made:

1. Changing the output inverter to a NAND/NOR will increase the parasitic delay of the output gate.
2. The precharge and keeper devices will be smaller, causing less parasitic loading and creating a faster circuit.
3. The precharge and keeper device area will be nearly identical.
4. Layout is more difficult because of the need to route two internal dynamic nodes.

If the speed loss due to the incremental delay of the NAND/NOR is less than the speed gain for the internal dynamic node loading reduction, then compounding will improve the speed of the gate.

Note that if one is constructing a wide AND gate by compounding two precharge NAND gates into a static NOR, one will need separate keeper–inverter pairs. This is to avoid short circuit current from flowing in the event that one stack evaluates and the other does not Figure 5.9.

5.8   Late Arriving Input on Top

Placing the late arriving input as the top signal in the evaluation stack will reduce the delay for two reasons:

1. The capacitance below that input is discharged early.
2. There is no body-effect-related slowing during evaluation.

Note that if one is unsure whether one input is early with respect to the other, one should model them as simultaneously switching in the simulation. Special note on the evaluation footer: Do not put the evaluation footer on top of the stack, even if it is the late input. This is especially true if using internal node precharge devices. These could cause short circuit current to flow if an input is at a logic one.

5.9   Making Keepers Weak

In cases where the keeper size works out to be smaller than the minimum width allowed by the process, one can do one of the following two things:

1. Increase the device length of the keeper to reduce its conductivity.
2. Use stacked PFET devices to reduce the overall conductivity (preferred).

If one decides to use stacked PFETs, one can choose to only drive the bottom device. The other devices would have their gates at V_{ss}. This is shown in Figure 5.10. The technique has some advantages and disadvantages as follows:

Advantages:
- Miller coupling to the dynamic node is reduced.
- Capacitive load on the output is reduced.

Disadvantages:
- Layout is more difficult because one has to route V_{ss} up to the keeper.
- Layout is more difficult because the two gates are connected to different signals.
- The evaluation delay of the gate may be a little greater.
- Negative bias temperature instability (NBTI) must be considered for the grounded device [36].

If using stacked PFET keepers, the inverter output should be connected only to the keeper nearest to the dynamic node. Keeper device width should be of

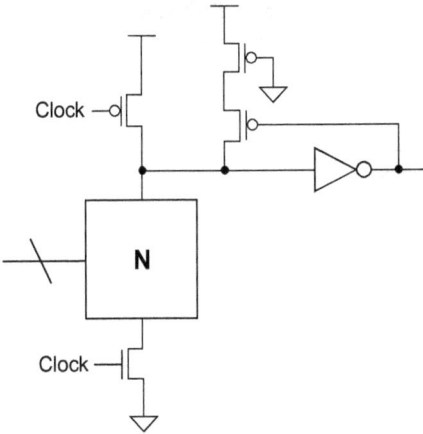

Figure 5.10 Stacked keepers.

minimum size. Keeper transistor length should be the same as in the evaluation NFET stack. Finally, keepers should be placed as close as possible to the receiver.

5.10 Conditional Keepers

If one is having trouble pulling down the dynamic net (writing the circuit), then one might have to resort to using a gated keeper. The circuit is shown in Figure 5.11. The operation of this circuit is fairly straightforward. During precharge (clock low), the keeper is off. When the clock goes high, the keeper stays off for a little bit. The delay time is equal to the delay through the buffer. Then the keeper turns on, if the NMOS tree did not pull down the dynamic node.

Another implementation of a conditional keeper [44] is shown in Figure 5.12.

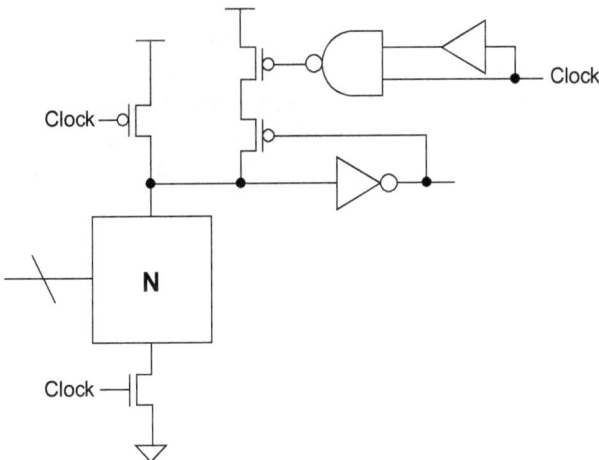

Figure 5.11 Conditional keeper.

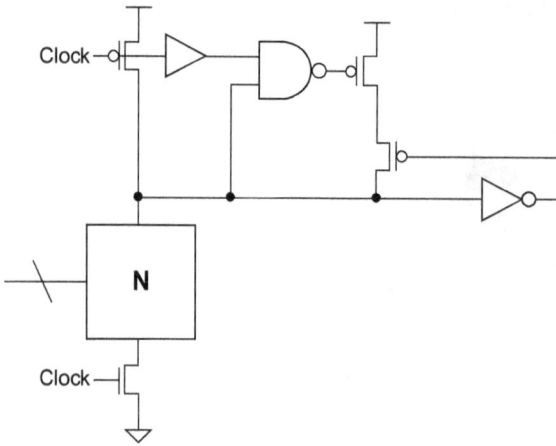

Figure 5.12 Another conditional keeper.

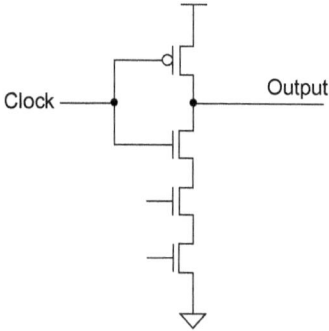

Figure 5.13 Poorly designed precharge circuit.

5.11   Placement of the Evaluation Device

As a general rule, do not place the evaluation device close to the dynamic node, as shown in Figure 5.13. This is because all the intermediate capacitances could be discharged during the precharge phase. Then, when the clock goes HIGH, severe charge sharing could take place [6]. Furthermore, if internal node prechargers are used, they could cause short circuit current to flow if an input is at a logic one.

6 Other Precharge Logic Styles

6.1 MODL

MODL [13] stands for "Multiple Output Domino Logic" and is a nice way to get more logic functionality out of standard domino gates. It is best to look at Figure 6.1 to visualize how multiple outputs are taken from a single gate.

What we have is the regular output F and an additional output F_1. F_1 is a subset of F, such that logic trees $F_1 \cap F_2 = F$. So the additional function is achieved by reusing a portion of an already existing logic tree.

MODL is particularly useful for blocks, such as carry generating circuits, which have a lot of recurrent logic [13]. For such circuits, device counts have been reduced up to a factor of two. So reduction in circuit area is one obvious advantage. Because this technique leads to fewer gates, the overall delays are reduced. Because there are fewer gates, there is also less power dissipation. The use of an additional precharge device to precharge the additional output becomes a must.

6.2 NORA Logic

Domino logic is not the only attempt at making precharge logic practical. NORA (NO Race) logic is another type, using both NMOS and PMOS logic trees [14,15]. But the NMOS and PMOS logic trees are used in alternating logic gates, as shown in Figure 6.2. This logic family does not need a static inverting gate at the output of every precharged gate, and so inversions are obtained easily.

But there are some problems. NAND gates still have series transistor stacks, which are slow. PMOS trees can get big because PMOS devices are inherently weaker, so PMOS trees could present a large load capacitance to the gates that drive them. The noise margins are small [4]. If the output of an NMOS stage drops below V_{tp}, then the following PMOS tree can turn on and evaluate. As stated in Ref. [4], do not use NORA logic, just say no!

There are some variations of NORA logic, such as ZIPPER logic [16]. In this case, precharge and predischarge devices of each stage never get turned off fully in order to reduce noise-sensitivity problems. Thus, these devices perform the same function as keepers in domino logic. Small currents maintain proper signal levels on the dynamic nodes, during evaluation phases. However, there is now additional complexity in clock generation as well as difficulty in maintaining appropriate leakage currents.

Synchronous Precharge Logic.

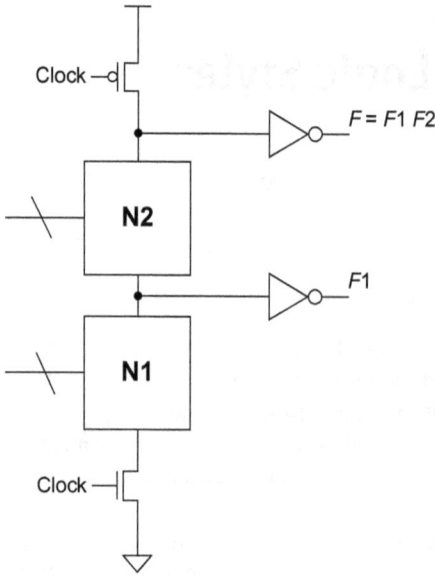

Figure 6.1 MODL circuit diagram.

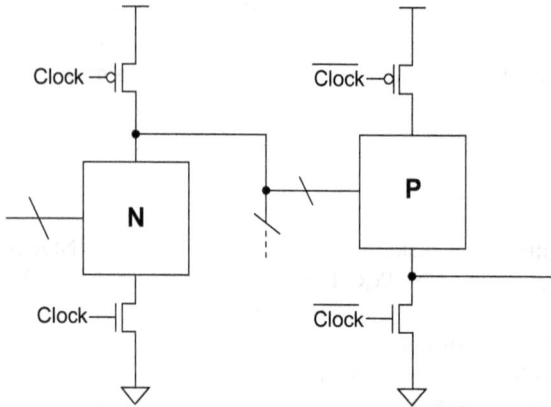

Figure 6.2 NORA logic.

6.3   Postcharge Logic

Postcharge logic, also known as self-resetting logic, has the advantage that it does not load down the clock, unlike precharge logic. The dynamic node is restored to a 1 logic level after the gate has evaluated [39]. The evaluation is what triggers this postcharge event. The name "postcharge" suggests a self-timed pipeline. Each stage is precharged again as soon as its data has been consumed by the next stage [4]. This is shown in Figure 6.3. When the dynamic node (x) falls, the output goes HIGH. Subsequently, after some delay, node (n) goes LOW to precharge the dynamic node back up to V_{dd}. Postcharge logic finds applications where a small percentage of gates switch in a cycle, such as a memory decoder.

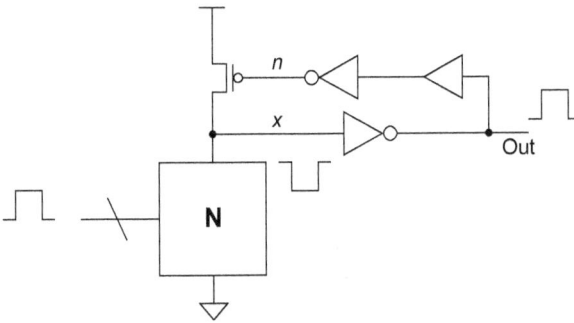

Figure 6.3 Postcharge logic gate.

The signal being propagated is buffered and used as the precharge or reset signal. By using a buffered form of the output, the output loading is kept almost as low as in clocked precharge logic while local generation of the reset ensures that it is properly timed and only occurs when needed [40].

So the output is fed back to the precharge control input and, after a specified time delay, the pull-up is reactivated. The delay line is implemented as a series of inverters. The signals that propagate through these circuits are pulses. The width of the pulses must be controlled carefully or else there may be contention between the NMOS and PMOS devices, or worse, oscillations may occur. The input to a postcharge gate needs to be a pulse, whose duration must be less than the reset delay. Thus, postcharge circuits are very fast and dense but are difficult to time. Inputs to the gate must arrive within a specific interval, made even more difficult by the high process sensitivity of the style.

So we need small pulses to avoid crowbar current during reset. We need to ensure that inputs do not come in before reset has completed. Otherwise, the input is delayed, so the output is delayed, and the next reset is delayed. This can become worse with every cycle, which is called "pulse walk out." Solutions are to back off on the operating frequency and/or make the reset path shorter. If the reset signal "r" is qualified by the input as shown in Figure 6.4, then we have additional gate delays in the reset path and any delay the input introduces. This is another type of

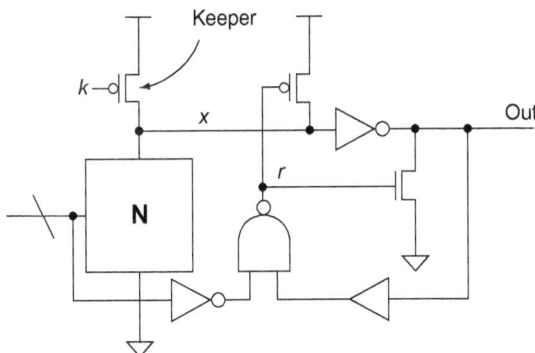

Figure 6.4 No evaluation device.

"pulse walk out." A solution is to keep the input pulses short, not to delay the reset.

A slightly more advanced postcharge logic gate is shown in Figure 6.5.

In this figure, we can see that an evaluation device and a keeper have been added. The evaluation protection is only necessary in the first stage of a postcharge logic chain. This prevents unwanted transitions much like in clocked precharge logic. Unlike precharge logic, however, postcharge logic circuits will recover from an unwanted transition. But this is slow and generally cannot be counted on for robust circuit performance. The evaluation protection forms a synchronization point. True asynchronous use can be achieved if no evaluation protection is used. Time borrowing is then possible. In this case, both the rise and fall input delays have to be optimized. The evaluation signal "e" can be generated in a few ways:

- It can be driven by the clock.
- It can be set to "r," which would turn off the evaluation only during reset.

The evaluation footer and signal "e" can be eliminated if:

- The inputs come from a monotonically rising pulse generator.
- Evaluation protection can also be achieved if we pull the inputs LOW with NOR gates controlled by the clock. This may be good if we have a high fan-in NAND stack.
- We can also feed a signal, which is a NAND of a signal from each parallel stack of transistors, back to the reset path, as shown in Figure 6.4.

The keeper signal "k" can be generated in a few ways:

- It can be a NAND of one signal from each parallel stack of transistors.
- It can be driven by the output "out" to form a classic keeper configuration.

Power dissipation is reduced in two ways. One of the advantages of self-resetting logic is that when the data present at evaluation does not require the dynamic node to discharge, the precharge device is not active, which reduces the power dissipation [41]. In addition, the clock infrastructure is now limited to the latches that launch and receive the signals, eliminating much of the clock wire and gate load, as compared to clocked precharge logic.

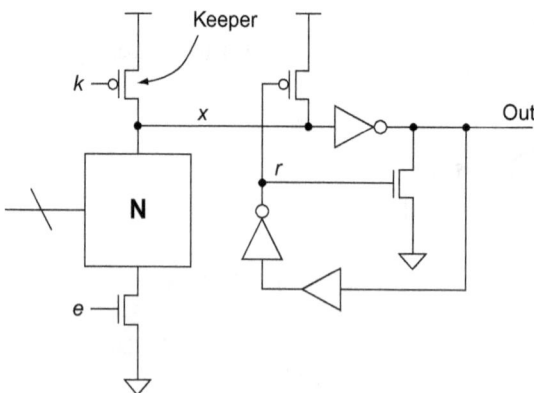

Figure 6.5 Improved postcharge gate.

6.4 CD Domino

CD domino stands for "clock delayed" domino. This is a self-timed logic style, which eliminates the fundamental monotonic signal requirement imposed on standard domino logic. It does this by propagating a clock signal in parallel to the logic, providing a dedicated clock to each logic stage [43]. A circuit diagram is shown in Figure 6.6 [45]. After the output inverter has fully developed its output, the delay circuit produces a delayed clock output that is provided to subsequent stages.

Because of the dedicated clocks, CD domino provides both inverting and noninverting functions. The output inverter is not essential because the clock will always arrive after the interval when evaluation should have been completed. Design margin must be added to the clock delay, shown in Figure 6.7, to guarantee function across process corners.

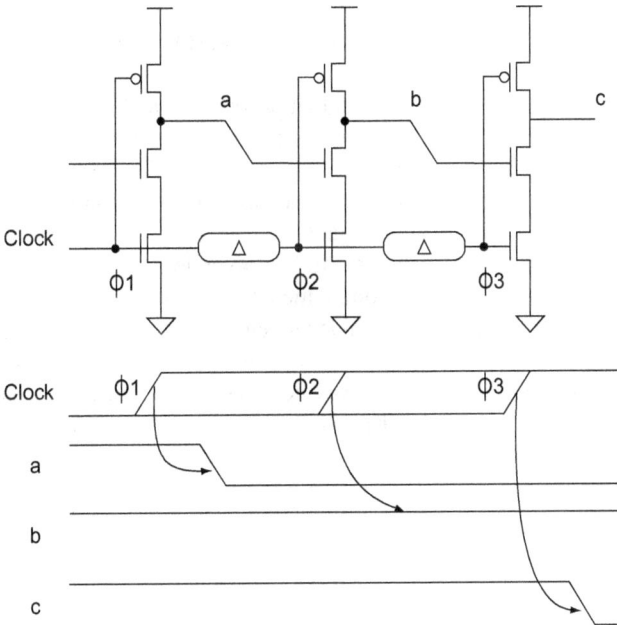

Figure 6.6 CD domino logic chain.

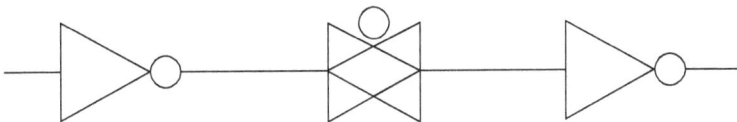

Figure 6.7 CD domino clock delay element.

6.5   NTP Logic

NTP logic stands for "noise tolerant precharge" logic. The precharge circuit's vulnerability to noise is associated with that feature that makes it fast. NTP logic provides higher noise immunity by avoiding floating dynamic nodes during evaluation [42]. The circuit is shown in Figure 6.8. The extra PFETs provide active noise immunity by providing a static type complementary logic pull-up. These devices are sized small.

NTP logic never allows a dynamic node to float. The complementary PFET logic is always on when the evaluate path to ground is off. These extra PFETs are not large enough to accomplish the precharge function, yet are sufficient to provide a high degree of noise immunity. In addition, the keeper is now not necessary and it is eliminated. The disadvantages are some cost in speed and area, as well as wiring complexity in layout.

6.6   Differential Cascode Voltage Switch Logic

Designers sometimes use domino gates that accept true and complementary monotonic inputs, and produce true and complementary monotonic outputs. Such gates are constructed with differential cascode voltage switch (DCVS) logic [2]. It is also known as dual-rail domino. Figures 6.9 and 6.10 show a dual-rail domino AND/NAND gate and also an XOR/XNOR gate.

Building dual-rail domino gates is similar to single rail design but requires the implementation of both true and complementary versions of functions. For some functions, these trees are completely independent, as shown in Figure 6.9. For other functions, the trees are partially shared as shown in Figure 6.10. Sharing reduces the input capacitance and so it improves both the speed and area. Sharing can be determined by inspection or by Karnaugh maps or tabular methods [32]. Dual

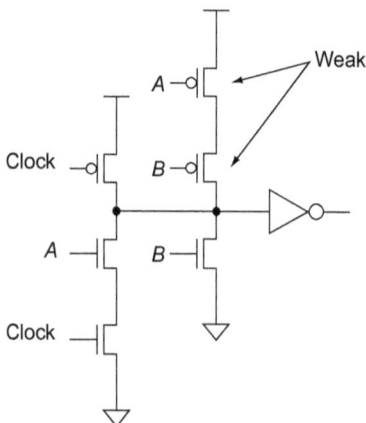

Figure 6.8 NTP logic gate.

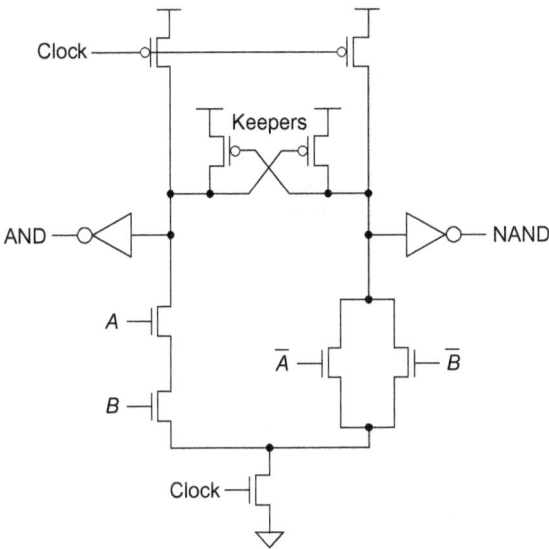

Figure 6.9 Dual-rail domino AND/NAND.

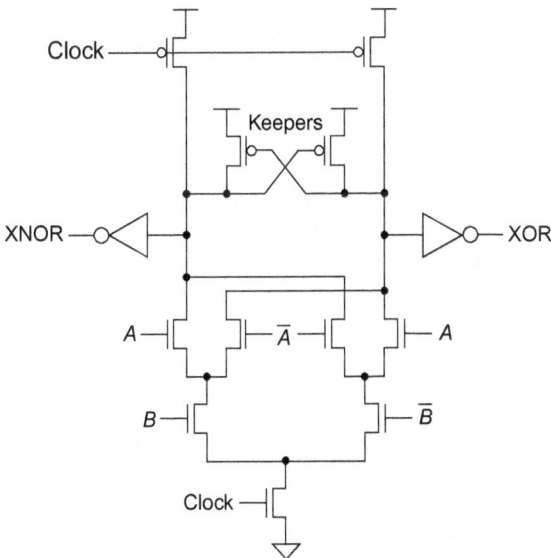

Figure 6.10 Dual-rail domino XOR/XNOR.

implies that parallel devices become series devices. This means that all dual-rail gates have series structures. This is one disadvantage of dual-rail gates, in that one cannot build a large fan-in gate. Another disadvantage of dual-rail domino is that it needs twice as many wires to carry the dual-rail signals and greater clock load.

By generating the monotonically rising true and complementary versions of all outputs, we can take advantage of one more thing. Each signal can indicate both value and completion [4] (see Table 6.1). The signal (on two wires) is not yet valid,

Table 6.1 DCVS Output Meaning

x_h	x_l	Meaning
0	0	Reset = Not yet evaluated
0	1	Ready with value FALSE
1	0	Ready with value TRUE
1	1	Not used = Never Occurs = Error

if both wires are LOW. When one wire is asserted, the signal has become stable. The key advantage is that blocks can be enabled before data arrives. This gets rid of the clock setup time.

In Figure 6.11, we can see how these output signals can be used to control the precharge and evaluation of the next stage. Here we use the dual-monotonic completion signal of a preceding gate as the precharge control signal of the next gate. This lowers the power consumption in every gate down the domino chain because the chain is now asynchronous (self-timed).

Here is a quick note about the cross-coupled keepers. Because the keepers do not need to be driven by the same side's inverter output, but rather from the complement's side output, this increases the noise margin. Also, in this way, the delay required to overcome the switchpoint of the keeper halflatch is avoided.

6.7 DCML

DCML stands for "dynamic current mode logic," and it is a differential small swing logic family. It is relatively fast because of the small swing. The small swing also reduces the dynamic power and cross talk. It has high noise immunity because it is a

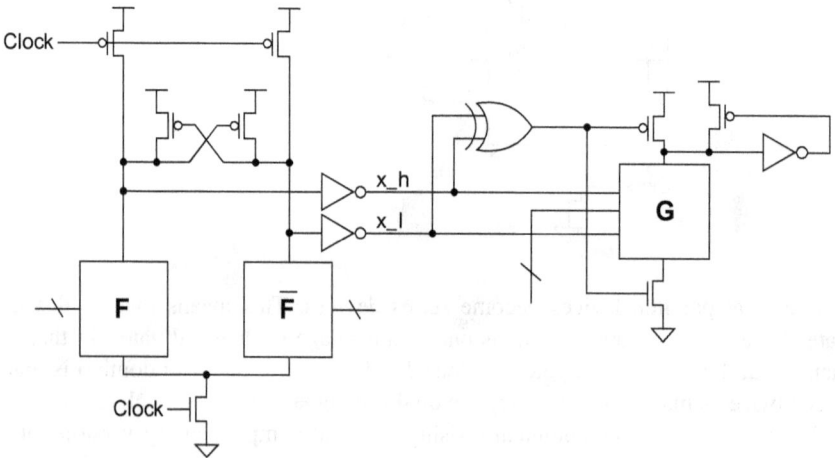

Figure 6.11 Dual-monotonic completion signal generation.

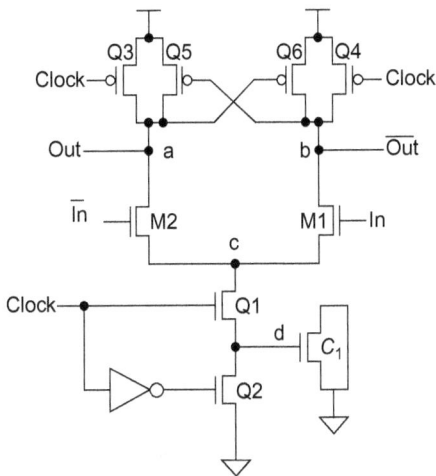

Figure 6.12 DCML inverter.

differential circuit. Generally, these are good for mixed signal applications. The circuit diagram is shown in Figure 6.12. Devices Q2, Q3, and Q4 handle the precharge and evaluation operations. Q1 and C1 form a dynamic current source. Q5 and Q6 form a latch to preserve the logic value and also act as keepers to offset leakage. Finally M1 and M2 are the logic transistors, an inverter in this simple case.

During precharge Q3 and Q4 precharge the outputs HIGH, while Q1 ensures there is no path to ground. During evaluation, the clock goes HIGH and Q1 turns on. Q2 turns off at the same time, so Q1 and C1 form a dynamic current source. The dynamic power of this circuit is small because the input capacitance can be small. There is no static power dissipation as Q1 and Q2 never turn on simultaneously. The waveforms for this circuit are shown in Figure 6.13.

Transistor C1 acts like a capacitor ($C_{gs} + C_{gd}$) and it limits the charge transferred from the output nodes. C1 should be sized considering the load capacitance (fanout) and required output swing. C1 was around 4–5% of the gate area in a 0.6 μm process. So the size is small, and it is largely because the voltage swing is small. Furthermore, C_{ox} is large because t_{ox} decreases with technology generations. Thus, the width of C1 need not be so large. Figure 6.14 shows the current source waveforms for this circuit. Q1 acts like a current source during evaluation. Node d starts to rise and Q1 turns off when $V_{ds} = 0$.

6.8 SOI Precharge Logic

Although silicon on insulator (SOI) devices have advantages in speed performance, there are circuit problems associated with them. Due to the floating body effect, the body-source voltage may have substantial drop, which generates considerable leakage current. Consequently, the dynamic node may be corrupted with the wrong state. To solve this problem, the circuit in Figure 6.15 has been used [38].

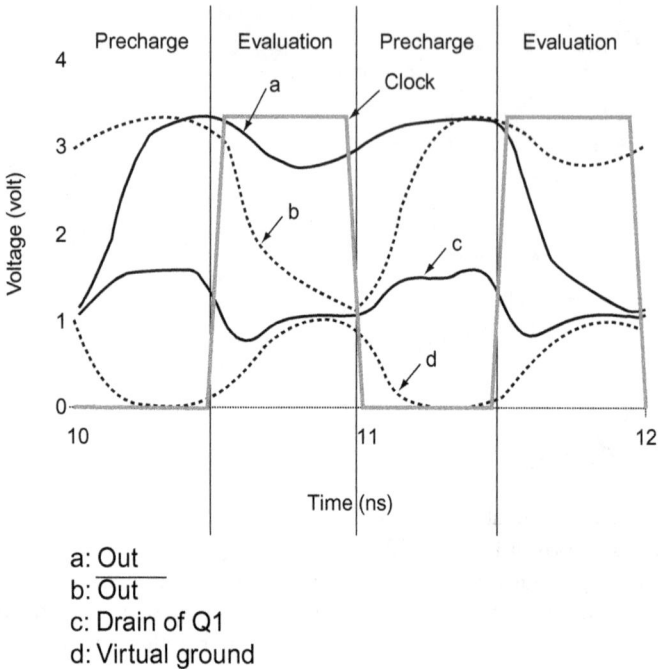

Figure 6.13 DCML voltage waveforms.

At the intermediate node, a predischarged PMOS device with its gate controlled by the clock has been added such that the charge accumulating in the body of the NMOS device above the intermediate node during the precharge phase can be discharged. Thus, the potential of the dynamic node does not drop to an erroneous logic state, due to a sudden drop from HIGH to LOW in the intermediate node during the evaluation phase. The coupling is eliminated. This circuit topology is suitable for SOI, which does not cause any charge-sharing problems due to the smaller junction capacitances, as compared to the bulk silicon counterpart.

6.9   Advanced Work

Advanced work on precharge logic has been done. Asynchronous precharge logic is a hot topic discussed in Refs. [4,17,18]. Asynchronous precharge logic relies on generating both the true and complement of a signal [4]. The two signals are then used as precharge control signals for the next logic gate. BiCMOS precharge logic free from charge sharing is discussed in Ref. [19]. Here, dynamic nodes are isolated from BJT devices to avoid leakage. Another possibility for BiCMOS precharge logic is discussed in Ref. [20], where the base of a BJT is precharged LOW (with the emitter tied to V_{ss}). Therefore, there is no leakage. Gallium arsenide (GaAs)

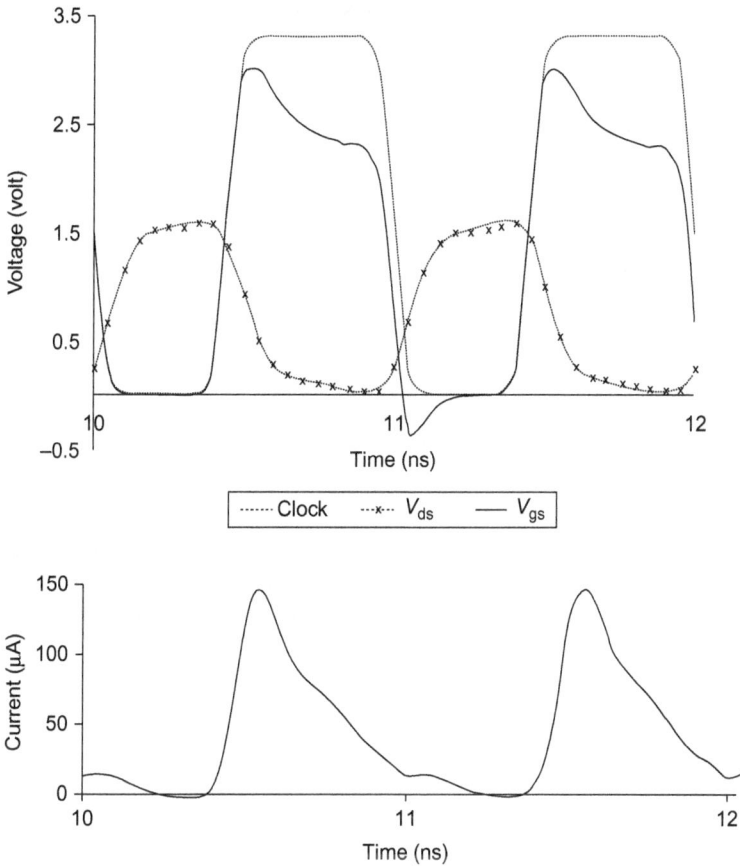

Figure 6.14 DCML current source waveforms.

precharge logic is discussed in Ref. [21]. MESFET leakage problems have been eliminated using coupling capacitors or high impedance (weak) transistors to replace lost charge.

Adiabatic (low power) precharge logic is discussed in Ref. [22]. Fast circuits have been designed with clock and data precharged dynamic (CDPD) logic [23]. Ternary precharge logic is discussed in Ref. [24]. This of course deals with four "state" logic as opposed to two state, such as binary. Finally, wave pipelining is discussed in Ref. [25]. This is a fascinating field where data is clocked into a combinatorial circuit with a clock phase smaller than the delay through the combinatorial logic. Two signals can be simultaneously propagating on the same logic path, separated only by a little bit of time. The signals are seen as waves.

As can be seen, there is a world of information available about precharge/dynamic logic. This book points out only some of the advanced research topics. Some of these are more practical and more common than others. We must also keep in mind that the industry is moving away from precharge logic since about

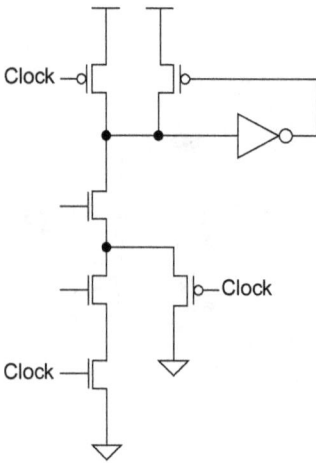

Figure 6.15 SOI precharge gate.

the 40−45 nm node. The reduced noise margin and reduced operation margin (due to reduced power supplies) are the major concerns. RAM designers continue to use precharge logic beyond these nodes in very controlled cases. However, the risk-versus-reward tradeoff and the requirement of more involved design and verification are making the use of precharge logic much rarer in 32/28/20 nm designs.

7 Clocked Set–Reset Latches

This chapter describes how a CSR latch works as well as the timing aspects related to these latches. The CSR latch is also known as the dynamic latch, zero keeper, glitch latch, set dominant latch, C^2MOS latch, and dynamic to static converter.

The monotonic properties of precharge circuits can be exploited to simplify the required latch on the output of such a path [29]. This is shown in Figure 7.1. The figure shows a chain of domino gates enabled by clock1, a latch, and the first stage of a clock2 domino chain. At the end of the evaluate phase, when clock1 falls, the value on signal B must be held steady while the clock1 chain precharges. The latch is simplified (compared to a standard four-transistor latch), by observing that at this point signal A is monotonically rising. Thus, the PFET in the latch can only turn off, and no gating PFET clocked by clock1 is needed. This concept is similar to the Svensson latch [28], with the monotonicity of domino logic eliminating the need for the first stage of the Svensson latch.

This latch avoids the hold-time problems associated with fast precharging, as discussed in Chapter 2. Because the output of the CSR latch is static, it functions as a dynamic to static converter. A CSR latch at the end of a precharge circuit block will ensure that the output of the precharge logic will hold its logic level through the next phase. Without the CSR latch, a race condition exists when the precharge circuit begins precharge because the precharge circuit will begin to propagate a zero, while the static logic in the next phase is attempting to propagate the earlier one.

Let us now look at a typical CSR latch. We know this type of latch is intended for a precharge type path such as a domino circuit or a bitline in a memory array. In these cases, we know that the output or bitline critical path will be a falling edge. Because the critical path through a CSR latch is simply through a NAND gate, a falling edge (0) will always produce a rising (1) output. This is significant because this property holds true, irrespective of the state of the latch (open or closed). Thus, we need not worry about paying for the clock uncertainty penalty. See Figures 7.2 and 7.3 for a basic CSR latch.

7.1 Memory Special Cases

Now we step back and look at the big picture. We need to see how these latches fit into the entire read path of a couple of memory special cases. See Figure 7.4, which shows two memory cases both with a read in the high phase of the clock.

Synchronous Precharge Logic.

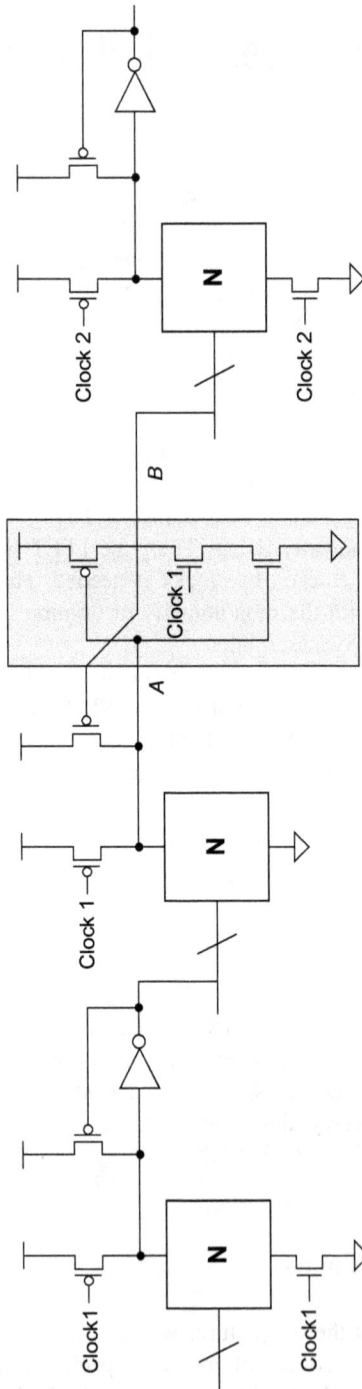

Figure 7.1 Domino logic with integral latch.

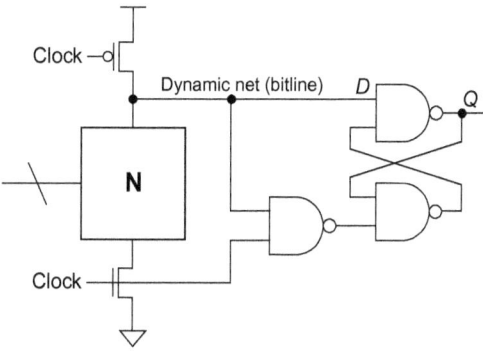

Figure 7.2 Basic CSR latch.

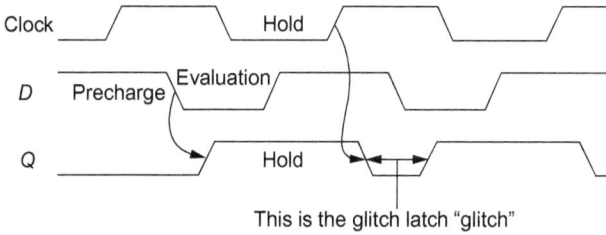

Figure 7.3 CSR latch timing.

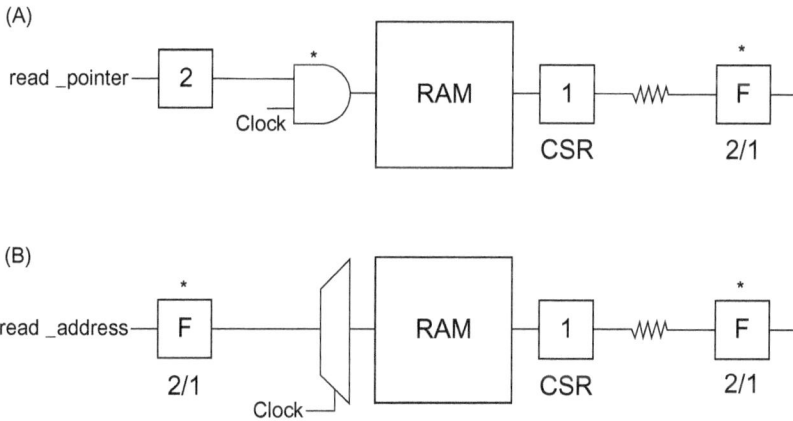

Figure 7.4 Memory cases: (A) fully decoded address (one hot) and (B) regular encoded address. *, sync point.

It can be seen that in the case of the encoded address, the high-phase CSR latch appears to be redundant. However, it is very much needed here because we must hold the output in the low phase, which is when the array precharges.

7.2 Building a CSR Latch

We start by taking a tri-state latch and removing the PMOS clock device, as shown in Figure 7.5. Right away we see that this satisfies our basic functionality requirement of passing a "0," irrespective of the state of the latch.

The placement of the clock device on the top node is good for noise in two ways. It prevents charge sharing and also lowers the noise coupling from D to Q. However, there are some unintended consequences. First, the capacitive load on the clock is highly state dependent, especially if the fan-out is large. Second, the LOW to HIGH transition is slower because it has to drag some of the capacitance of the stacked node.

Subsequently, we change the keeper loop from the cross-coupled inverters to the NAND, as shown in Figure 7.6. The NFET clock device becomes the NFET keeper.

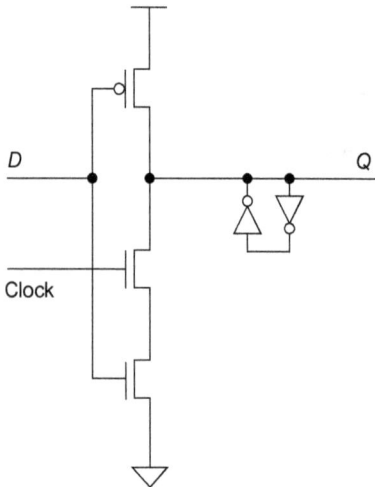

Figure 7.5 Basic dynamic latch.

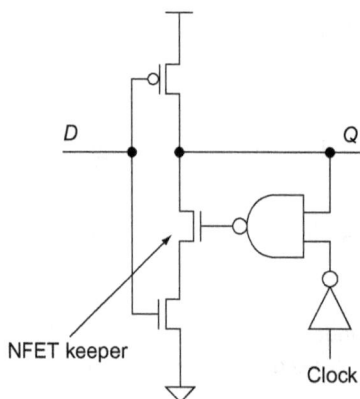

Figure 7.6 Dynamic latch with NAND loop.

Next, we add a PFET keeper (Figure 7.7).

Now we redraw the latch as two cross-coupled NANDs: (Figure 7.8).

Finally, we add a *third NAND gate* (Figure 7.9) so that nets RB and QB become a function of the input *D*. The problem is that the clock to the latch can go low late. If the latch clock goes LOW late, it can put the latch into "holding" mode after the input started to precharge. This means we would have missed the negative pulse on input *D*. This could happen, especially when one considers all *clock uncertainty*.

In the circuit below, *D* can put the latch into "holding" mode before the clock falls. We still need the latch clock to fall before *D* starts to precharge, but it is much easier to meet this hold-time margin. The clock does not need to propagate

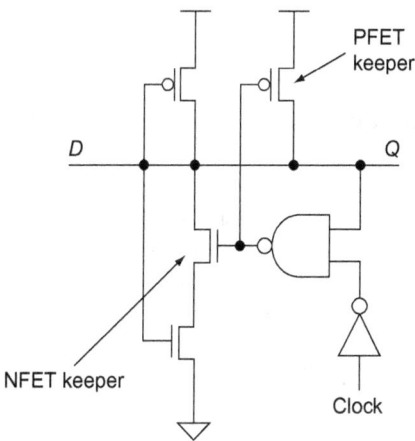

Figure 7.7 Adding a PFET keeper.

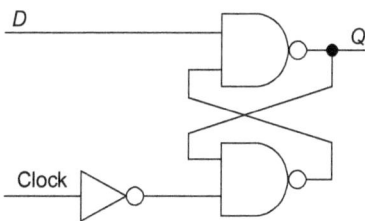

Figure 7.8 Redrawn as CSR latch.

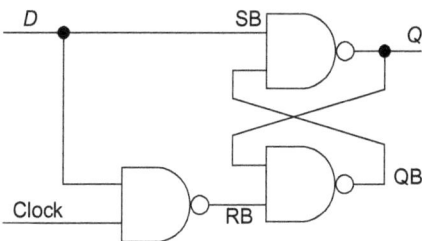

Figure 7.9 Adding NAND for hold time.

Table 7.1 SR Latch Truth Table

QB	Q	SB	RB
1	1	0	0
0	1	0	1
1	0	1	0
HOLD	HOLD	1	1

through the two NAND gates to get to QB, as QB is already low. See the truth table (Table 7.1).

7.3   Time Borrowing

If the circuit allows time borrowing from the next phase or cycle only, it is called *forward time borrowing*. As shown in Figure 7.10, phase 1 is borrowed into phase 2.

On the other hand, if the circuit allows time borrowing from the previous phase or cycle only, it is called *backward time borrowing*. This is not common.

The final type of time borrowing is *bidirectional time borrowing*. An example of bidirectional time borrowing is a CSR latch. This type of circuit allows both forward and backward time borrowing (Figure 7.11). If the delay of the circuit in phase 1 is

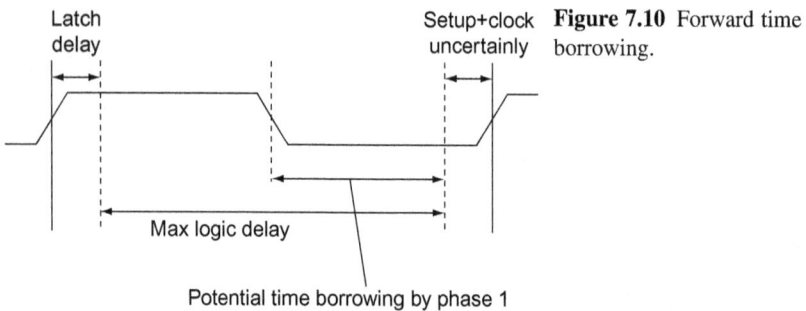

Figure 7.10 Forward time borrowing.

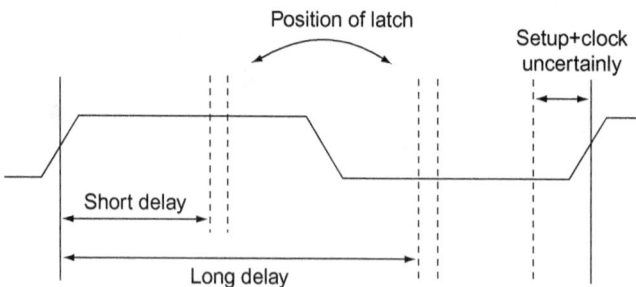

Figure 7.11 CSR latch time borrowing.

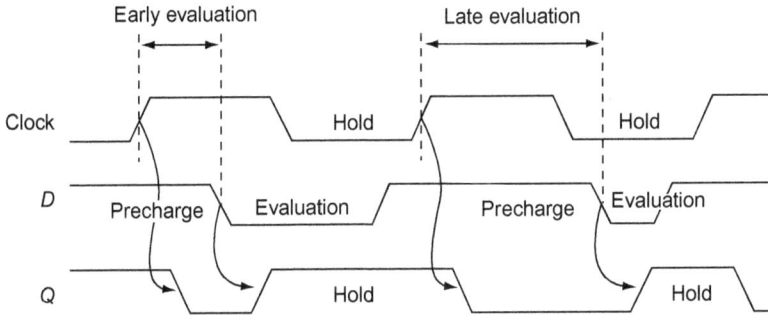

Figure 7.12 Early and late evaluation.

short, phase 2 can borrow backward into phase 1. On the other hand, if the phase 1 circuit takes longer than a phase, it can forward time borrow into phase 2.

Finally, Figure 7.12 shows the timing of the CSR latch with both an early and a late evaluation.

Note the *short evaluation pulse* when we have the late evaluation. This leads us into the Section 7.4, where we check that the latch still functions properly, despite the late evaluation. In the next section, we look at *hold-time margins*.

7.4 Hold-Time Margins

7.4.1 Margin 1

Margin 1 states that we must have net QB flip low before the *D* input precharges high (Figure 7.13).

This margin applies to the CSR latch shown in Figure 7.9. The *third NAND gate* allows net QB to be a function of the input *D*. This makes the margin into a frequency-dependent margin. The equation below resembles that of a setup time and we can view this as QB setting up to *D* rising. However, it is really the input *D* holding low past QB falling about which we are concerned.

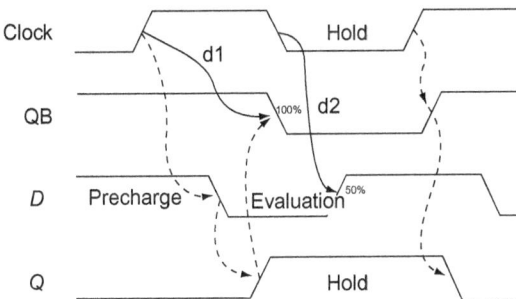

Figure 7.13 Hold margin 1.

Type: functional, frequency-dependent margin
Short path = D_1: QB fall
Long path = D_2: D rise
Margin calculation: Margin (ps) = D_2 + Tcyc/2 − D_1 − skew − jitter
Minimum margin: 0 ps

7.4.2 Margin 2

Margin 2 states that the clock-like reset net RB must go high (holding) before the D input precharges high. If the clock to the CSR latch goes low late, it can put the latch into "holding" mode after the input started to precharge. This means we would have missed the negative pulse on input D (Figure 7.14).

Type: functional, self-timed margin
Short path = D_1: RB rise
Long path = D_2: D rise
Margin calculation: Margin (%) = $(D_2 − D_1 − \text{skew})/(D_2 + D_1)$
Minimum margin: 20% (typical)
Where $D_2 + D_1$ is the loop delay.

7.5 Mintime

A CSR latch is remarkable in that the falling edge can pass through after the latch has closed. This is the *late evaluation* discussed in Sections 7.4.1 and 7.4.2. This pushes the latch further into the next clock phase. If one pushes the latch too much, one might have a mintime problem between the CSR latch and the latch/flop that follows.

This is a concern primarily for the reset edge of the CSR latch. The quick reset nature of this latch makes this edge worse. If one runs into this problem and does not want to hurt the critical timing, simply add delay on the reset path of the CSR latch. The more delay added, the more process, voltage and temperature (PVT) variation one adds, so keep that in mind. Run mintime margin simulations to make sure all is satisfactory (Figure 7.15).

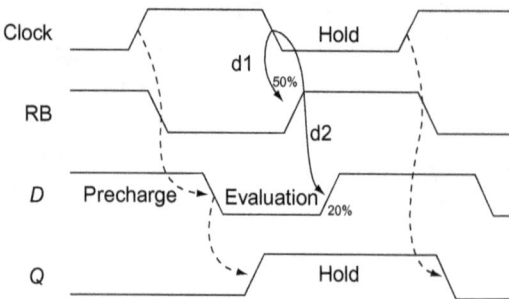

Figure 7.14 Hold margin 2.

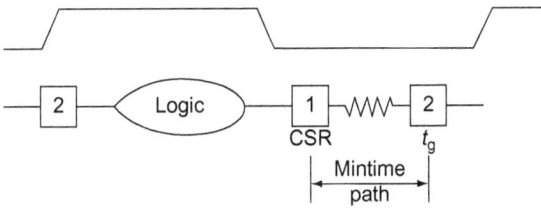

Figure 7.15 Mintime.

7.6 Alternative Topology

Here is a topology alternative to the CSR latch discussed in Section 7.2 (Figure 7.16).

7.7 The Other Phase

Till now we were focusing on phase-1 CSR latches. However, it is possible that a need arises for a phase-2 CSR latch. To do this, we simply invert the clock, as can be seen in Figures 7.17 and 7.18.

7.8 Two-Input Latch

Sometimes it is convenient to do a little bit of logic in the CSR latch. As such, we have a two-input CSR latch as shown in Figure 7.19.

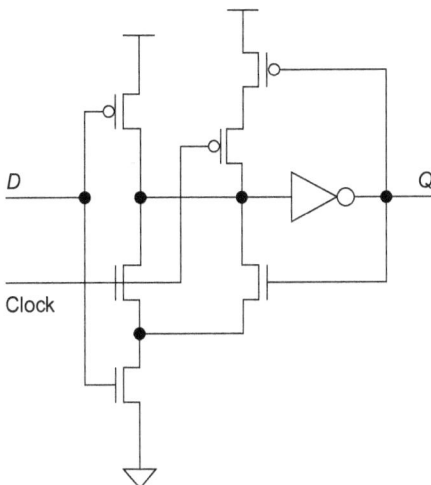

Figure 7.16 Alternative topology.

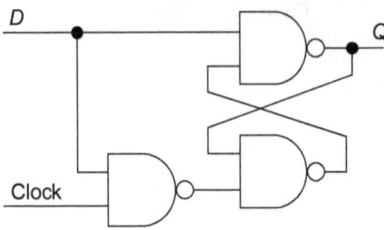

Figure 7.17 Phase-1 CSR latch.

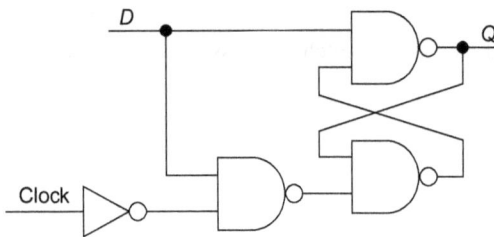

Figure 7.18 Phase-2 CSR latch.

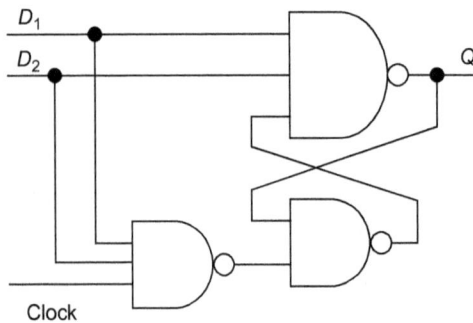

Figure 7.19 Two-input CSR latch.

7.9 Adding Scan

This is a scannable phase-1 CSR latch [31]. Logically, it is equivalent to our standard CSR latch, as can be seen in Figure 7.20.

The latch has to work with the following testing methodology: There is only one clock to handle both normal and scan functions, selected with a "scan enable" signal. The scan input is separate from the data input, while the scan output is a buffered version of the data output. The scan chain elements consist of a phase-2 latch followed by a phase-1 latch.

The circuit introduces a "dynamic load stack" to capture the dynamic data from the input. Because the static NAND loop is removed, a "D type" phase-2 and phase-1 latches are added to allow for scanning, as shown in Figure 7.20. The holding function for the CSR latch is now done by the "holding loop," which is

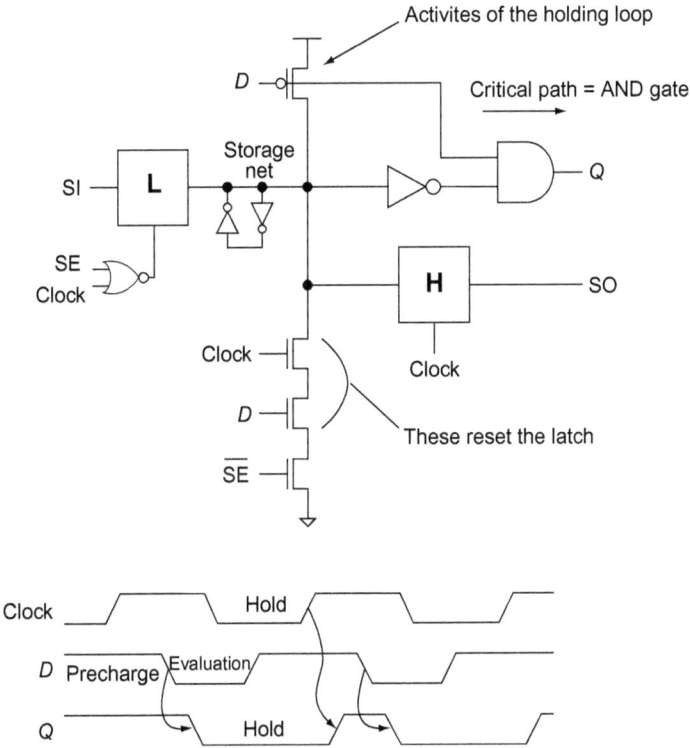

Figure 7.20 CSR latch with scan.

normally part of the phase-2 latch. This loop is active during normal functional operation. The loop is activated by the PMOS device. It is cleared (reset) by the dynamic load stack. The assumption is that the input "*D*" will be precharged HIGH during scanning.

8 Layout Considerations

Traditionally, precharged gates could be optimized to minimize delay. The pull-down delay could be reduced by progressively reducing the width of transistors moving up a stack [9]. By keeping the total device area constant and sizing each device to be 25% smaller than the one below it, a 20% reduction in delay can be achieved for a six-input NAND gates [9]. Even if the area is not preserved (we keep the bottom device the same size and simply move up the stack, reducing the size of each device), there can be some reduction in delay. Is should be noted that in submicron technologies, the reduction in delay is only 2−4% [10].

Figure 8.1 shows an example layout with such "tapered" devices. Looking at this layout, it is obvious that the spacing between poly lines is not minimum. This is due to the fact that in most technologies, diffusion must overhang the transistor by some amount X. A DRC violation is present if the diffusion to the right of the poly is made any smaller. Clearly, this increases both the capacitance and resistance of the drain/source regions, which then reduces the advantages of tapering.

Looking now at an equation will help explain why the delay is being reduced. Figure 8.2 represents a precharged gate as an RC network. Equation (8.1) is a representation of the Elmore delay for the circuit shown in Figure 8.2.

$$T_d = R_n C_1 + (R_n + R_a) + (R_n + R_a + R_b)C_1 + (R_n + R_a + R_b + R_c)C_0 \qquad (8.1)$$

$$= R_c C_0 + R_b(C_1 + C_0) + R_a(C_2 + C_1 + C_0) + R_n(C_3 + C_2 + C_1 + C_0) \qquad (8.2)$$

We can see from Eq. (8.2) that smaller resistances are multiplied by larger capacitive factors. Thus, the total delay is reduced. Another way to look at this is to take the top transistor (R_c) and lump all the other ones into a single resistance R_l. Since $R_c \ll R_l$, the effect of increasing R_c is minimal. But capacitance C_0 and C_3 are smaller, yet they discharge through more or less the same total resistance of the stack. This means that the RC products are smaller.

An efficient scaling procedure for series transistor stacks is discussed in Ref. [11]. This is mostly based on Monte Carlo simulations. A trial-and-error size selection is repeated several times to determine the best scaling.

Modern processes make it more difficult to do precharged style design. For example, restrictions are placed on the width and length that a device could have. This leads to limited legging choices, which in turn makes it difficult to accurately design circuits like keepers. However, the restrictions do make sense, as the more combinations that we have, the more variation we will get. Variation especially

Synchronous Precharge Logic.

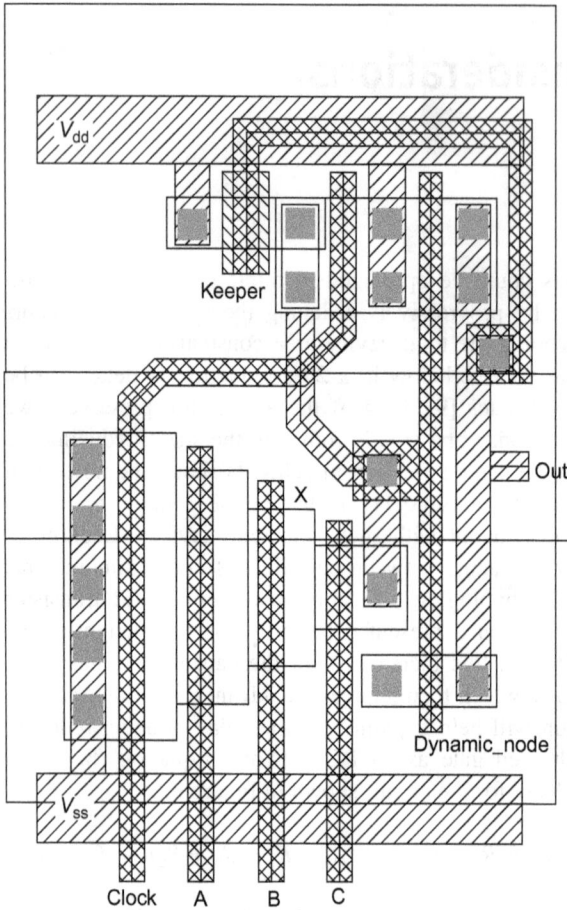

Figure 8.1 Tapering
transistors in a three-input
NAND gate.

increases with higher gate length. Fewer width/length combinations will have a higher chance of good yield and predictable behavior.

Large devices are not recommended because they have large gate resistance. Also, one gate contact driving through long gates is bad for high speed. Multiple voltage thresholds allow for trading off performance for leakage. But multiple V_ts require multiple mask layers. Higher V_t means higher doping and higher dopant variation. In addition, higher doping negatively impacts the I–V curve such that V_{dd} versus frequency is worsened. Higher V_t also is worse for negative bias temperature instability (NBTI) shift, which makes P versus N mismatch worse. L-shaped diffusion is not allowed in many cases, which makes tapering impossible. It is also best to minimize unique well shapes.

Let us now consider some common rules that designers should follow when doing precharge gate layout in modern processes:

• For process variation concerns related to the poly mask, it is recommended to keep the keeper device length equal to the NFET stack lengths.

R_c

R_b

R_a

R_n

C_0

C_3

C_2

C_1

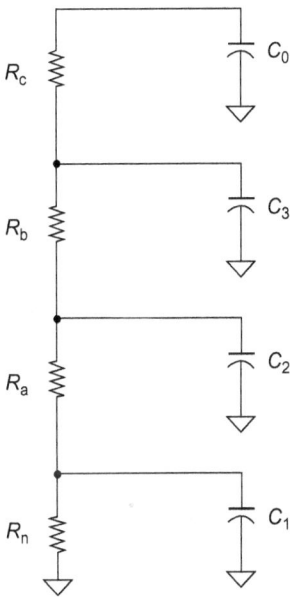

Figure 8.2 Stack of transistors represented by an RC network.

- Keepers should be placed as close as possible to the receiver.
- Keep dynamic nodes away from large drivers, especially if decoupling capacitors cannot be placed in the vicinity.
- All dynamic nodes are recommended to be covered by a V_{dd}/V_{ss} metal shield.
- All polysilicon must be shielded with dummy poly.
- Keeper layout should have extra space between poly edge and contact to allow for future upsizing.
- If space allows, include spare keepers with metal options.

Appendix: Logical Effort

The method of logical effort is a straightforward technique used to estimate delay in a CMOS circuit. Used properly, it can aid in selection of gates for a given function (including the number of stages necessary) and sizing gates to achieve the minimum delay possible for a circuit.

A.1 Derivation of Delay in a Logic Gate

Delay is expressed in terms of a basic delay unit, $\tau = 3RC$, the delay of an inverter driving an identical inverter with no parasitic capacitance; the unitless number associated with this is known as the normalized delay. The absolute delay is then simply defined as the product of the normalized delay of the gate, d, and τ:

$$d_{abs} = d\tau$$

In modern 45 nm processes, the absolute delay is approximately 4−5 ps.

The normalized delay in a logic gate can be expressed as a summation of two factors: *parasitic delay*, p (which is an intrinsic delay of the gate and can be found by considering the gate driving no load), and *stage effort*, f (which is dependent on the load as described below). The relationship between them is as follows:

$$d = f + p$$

The stage effort is divided into two components: a logical effort, g, which is the ratio of the input capacitance of a gate to that of an inverter capable of delivering the same output current, and an electrical effort, h, which is the ratio of the input capacitance of the load to that of the gate. The stage effort is then simply:

$$f = gh$$

Combining these equations yields a basic equation that models the normalized delay of a single logic gate:

$$d = gh + p$$

A.2   The Logical Effort of a Single Stage

The logical effort again is the ratio of the input capacitance of a gate to that of an inverter capable of delivering the same output current. Figure A1 shows the logical effort g of precharge logic gates compared to an inverter with a beta = 2. The figure also shows the parasitic delay p. Unfooted precharge gates have the lowest logical effort. Footed gates can approach the logical effort of unfooted gates at the expense of a larger clocked transistor. This increases the clock loading, which increases power consumption.

A.3   Multistage Networks

The total normalized path delay D can be expressed in terms of an overall path effort, F, and the path parasitic delay P (which is the sum of the individual parasitic delays):

$$D = F + P$$

The path effort is expressed in terms of the path logical effort G (the product of the individual logical efforts of the gates), and the path electrical effort H (the ratio

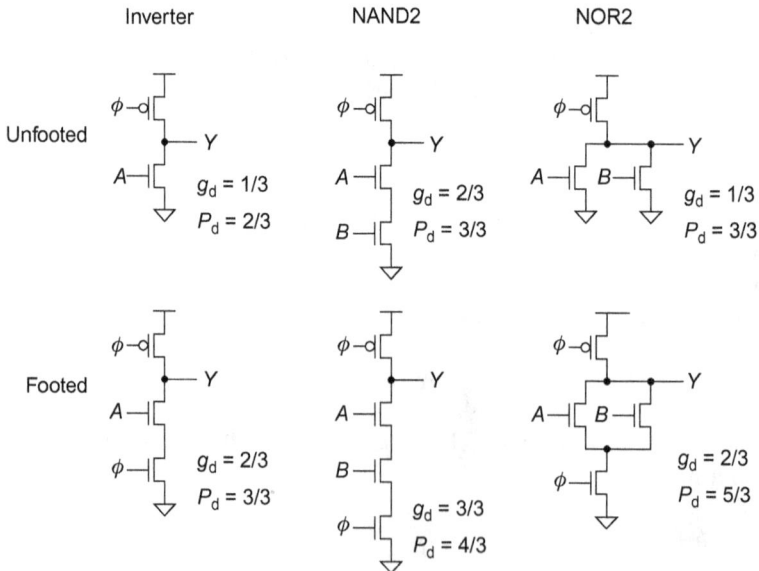

Figure A1 Logical effort in precharge logic.

of the load of the path to its input capacitance). For paths where each gate drives only one additional gate:

$$F = GH$$

However, for circuits that branch, an additional branching effort, b, needs to be taken into account. It is the ratio of total capacitance being driven by the gate to the capacitance on the path of interest:

$$b = \frac{C_{\text{onpath}} + C_{\text{offpath}}}{C_{\text{onpath}}}$$

This yields a path branching effort B, which is the product of the individual stage branching efforts. The total path effort is then:

$$F = BGH$$

The total delay for the path is given by:

$$D = BGH + P$$

A.4 Minimum Delay

It can be shown that in multistage logic networks, the minimum possible delay along a particular path can be achieved by designing the circuit such that the stage logical efforts are equal. For a given combination of gates and a known load, B, G, and H are all fixed, causing F to be fixed. Hence for N gates, the individual gates should be sized such that the individual stage efforts are:

$$f_{\text{min}} = F^{1/N}$$

Thus, the minimum delay of an N stage path is:

$$D_{\text{min}} = NF^{1/N} + P$$

How wide should the gates be for least delay? Working backward, apply capacitance transformation to find the input capacitance of each gate (i), given the load it drives:

$$f_{\text{min}} = gh = gC_{\text{out}}/C_{\text{in}}$$

$$C_{\text{in}(i)} = g_{(i)}C_{\text{out}(i)}/f_{\text{min}}$$

A.5   Best Number of Stages

Minimizing the number of stages is not always fastest. The ideal value for N can be found by calculating the best stage effort:

$$\mathrm{d}D/\mathrm{d}N = -F^{1/N} \ln F^{1/N} + F^{1/N} + p_{\text{inv}} = 0$$

Define the best stage effort:

$$p = F^{1/N}$$

or

$$p_{\text{inv}} + p(1 - \ln p) = 0$$

Then,

$$N = \log_4 F$$

References

[1] R. Krambeck, C. Lee, H. Law, High speed compact circuits with CMOS, IEEE JSSC SC-17 (3) (1982) 614−619.

[2] P. Gronowski, B. Bowhill, Dynamic logic and latches—part II, in: Proceedings VLSI, Circuits Workshop (VLSI Symposium), June 1996.

[3] Kernhof, Selzer, Beunder, Hoefflinger, Laquai, Schindler, Mixed static and domino logic on the CMOS gate forest, IEEE JSSC 25 (2) (1990) 396−402.

[4] T. Williams, Dynamic logic: clocked and asynchronous, in: IEEE International Solid State Circuits Conference Tutorial, 1996, pp. 1−24.

[5] Svensson, Yuan, High speed CMOS circuit technique, IEEE JSSC 24 (1) (1989) 62−70.

[6] Weste, Eshraghian, Principles of CMOS VLSI Design, second ed., Addison-Wesley, 1993, pp. 290−291, 301−303, 308−311.

[7] H. Bakoglu, Circuits, Interconnections, and Packaging for VLSI, Addison-Wesley, 1990, pp.289−303, 312−320.

[8] Larsson, Svensson, Noise in digital dynamic CMOS circuits, IEEE JSSC 29 (6) (1994) 655−662.

[9] M. Shoji, FET scaling in domino CMOS gates, IEEE JSSC SC-20 (5) (1985) 1067−1071.

[10] Hoppe, Nevendorf, Schmitt-Landsiedel, Specks, Optimization of high speed CMOS logic circuits with analytical models for signal delay, chip area, and dynamic power dissipation, IEEE Trans. Comput. Aided Des. 9 (3) (1990) 236−247.

[11] L. Wurtz, An efficient scaling procedure for domino CMOS logic, IEEE JSSC 28 (9) (1993) 979−982.

[12] D. Radhakrishnan, Design of CMOS circuits, IEE Proc. G 138 (1) (1991) 83−90.

[13] I. Hwang, A. Fisher, Ultrafast compact 32 bit CMOS adders in multiple output domino logic, IEEE JSSC 24 (2) (1989).

[14] N. Gonclaves, H. DeMan, NORA: a racefree dynamic CMOS technique for pipelined logic structures, IEEE JSSC SC-18 (3) (1983) 261−263.

[15] Renshaw, C. Lau, Race free clocking of CMOS pipelines using a single global clock, IEEE JSSC 25 (3) (1990) 766−769.

[16] Lee, E. Sheto, Zipper CMOS, IEEE Circuits Syst. Mag. (1986) 10−16.

[17] A. McAuley, Dynamic asynchronous logic for high speed CMOS systems, IEEE JSSC 27 (3) (1992) 382−388.

[18] C. Farnsworth, D. Edwards, S. Sikand, Utilizing Dynamic Logic for Low Power Consumption in Asynchronous Circuits, The University of Manchester, Manchester.

[19] J. Kuo, C. Chiang, Charge sharing problems and the dynamic logic circuits: BiCMOS versus CMOS and a 1.5 V BiCMOS dynamic logic circuit free from charge sharing problems, IEEE Trans. Circuits Syst.—1: Fundam. Theory Appl. 42 (11) (1995) 974−977.

[20] S. Menon, A. Jayasumana,Y. Malaiya, A Novel High Speed BiCMOS Domino Logic Family, Colorado State University, Fort Collins, CO, pp. 21−24.

[21] D. Hoe, C. Salama, Pipelining of GAAS Dynamic Logic Circuits, University of Toronto, Toronto, ON, pp. 208–211.

[22] A. Dickinson, J. Denker, Adiabatic Dynamic Logic, AT&T Bell Labs, Holmdel, NJ, pp. 282–285.

[23] H. Lindkvist, P. Anderson, Dynamic CMOS Circuit Techniques for Delay and Power Reduction in Parallel Adders, Lund University, Sweden, pp. 121–130.

[24] J. Wang, C. Wu, M. Tsai, Low power dynamic ternary logic, IEE Proc. G 135 (Pt. G, 6) (1988) 221–230.

[25] W. Lien, W. Burleson, Wave Domino Logic: Timing, Analysis and Applications, University of Massachusetts, Amherst, MA, pp. 2929–2952.

[26] D.-O. Lee, et al., Mist deposited high-k dielectrics for next generation MOS gates, Solid State Electron. 46 (2002) 1671–1677.

[27] I.E. Sutherland, R.F. Sproull, Logical effort: designing for speed on the back of an envelope, in: Advanced Research in VLSI, ARVLS'I91, Santa Cruz, 1991.

[28] M. Afghahi, C. Svensson, A unified single-phase clocking scheme for VLSI systems, IEEE JSSC 25 (1) (1990) 225–233.

[29] D.W. Bailey, B.J. Benschneider, Clocking design and analysis for a 600-MHz alpha microprocessor, IEEE JSSC 33 (11) (1998).

[30] D. Harris, M. Horowitz, Skew-tolerant domino circuits, IEEE JSSC 32 (1) (1997) 1702–1711.

[31] M. Smoszna, A scannable, clocked SR latch—for a single functional/scan clock system, US Patent application at Nvidia Corporation, Docket number SC11-0204-ISF, 2011.

[32] K. Chu, D. Pulfrey, Design procedures for differential cascode voltage switch circuits, IEEE JSSC SC-21 (6) (1986) 1082–1087.

[33] K. Nowka, T. Galambos, Circuit design techniques for a gigahertz integer microprocessor, in: Proceedings of International Conference on Computer Design, October 1998, pp. 11–16.

[34] C. Heikes, G. Colon-Bonet, A dual floating point coprocessor with an FMAC architecture, ISSCC Digest of Technical Papers, February 1996, pp. 354–355.

[35] IBM J. Res. Develop. 40 (1) (1996) (special issue on soft errors).

[36] D. Greenhill, Robust circuit design, flawed circuit design, in: IEEE International Solid State Circuits Conference Tutorial, 2005, pp. 14–30.

[37] P. Gronowski, et al., High performance microprocessor design, IEEE JSSC 33 (5) (1998) 676–685.

[38] A.G. Aipperspach, et al., A 0.2 μm 1.8 V SOI 550 MHz 64-b power PC microprocessor with copper interconnects, IEEE JSSC 34 (11) (1999) 1430–1435.

[39] R. Heald, J. Holst, A 6 ns Cycle 256-kb cache memory and memory management unit, IEEE JSSC 28 (11) (1993) 1078–1083.

[40] R.A. Haring, M.S. Milshtein, T.I. Chappell, S.H. Dong, B.A. Chapell, Self resetting logic and incrementer, in: Proceedings of the IEEE International Symposium on VLSI Circuits, 1996, pp. 18–19.

[41] G. Yee, C. Sechen, Clock-delayed domino for adder and combinational logic design, in: Proceedings of the IEEE/ACM International Conference on Computer Design, October 1996, pp. 332–337.

[42] Yamada, et al., A 13.3 ns double-precision floating point ALU and multiplier, in: Proceedings of the 1995 ICCD, 1995, pp. 466–470.

[43] G. Yee, et al., Clock-delayed domino for adder and combinational logic design, in: Proceedings of the 1996 ICCD, 1996, pp. 332–337.

[44] K.-I. Oh, L.-S. Kim, A clock delayed sleep mode domino logic for wide dynamic or gate, in: International Symposium on Low Power Electronics and Design, August 2003, pp. 176−179.

[45] B.A. Shinkre, J.E. Stine, A pipelined clock-delayed domino carry-lookahead adder, in: Great Lakes Symposium on VLSI, April 2003, pp. 171−175.

[46] Narendra, Borkar, De, Antoniadis, Chandrakasan, Scaling of stack effect and its application for leakage reduction, in: International Symposium on Low Power Electronics and Design, August 6−7, 2001, pp. 195−200.

www.ingramcontent.com/pod-product-compliance
Lightning Source LLC
Chambersburg PA
CBHW060322220326
41598CB00027B/4393